Delmar's Test Prepa

Medium/Heavy Truck Test

Diesel Engines (Test T2)

Technical Advisor
John F. Kershaw

Africa • Australia • Canada • Denmark • Japan • Mexico • New Zealand • Philippines
Puerto Rico • Singapore • Spain • United Kingdom • United States

NOTICE TO THE READER

Publisher does not warrant or guarantee any of the products described herein or perform any independent analysis in connection with any of the product information contained herein. Publisher does not assume, and expressly disclaims, any obligation to obtain and include information other than that provided to it by the manufacturer.

The reader is expressly warned to consider and adopt all safety precautions that might be indicated by the activities herein and to avoid all potential hazards. By following the instructions contained herein, the reader willingly assumes all risks in connection with such instructions.

The Publisher makes no representation or warranties of any kind, including but not limited to, the warranties of fitness for particular purpose or merchantability, nor are any such representations implied with respect to the material set forth herein, and the publisher takes no responsibility with respect to such material. The publisher shall not be liable for any special, consequential, or exemplary damages resulting, in whole or part, from the readers' use of, or reliance upon, this material.

Delmar Staff:
Business Unit Director: Alar Elken
Product Development Manager: Jack Erjavec
Executive Marketing Manager: Maura Theriault
Channel Manager: Mona Caron
Executive Production Manager: Mary Ellen Black
Cover Design: Paul Roseneck
Cover Image: © 1998 Corbis Corp.

Printed in Canada
4 5 6 7 8 9 10 XXX 05 04 03 02 01

For more information, contact Delmar, 3 Columbia Circle, PO Box 15015, Albany, NY 12212-0515; or find us on the World Wide Web at http://www.delmar.com

ISBN 0-7668-0560-3

Contents

Section 1 The History of ASE

Section 2 Take and Pass Every ASE Test

Section 3 Are You Sure You're Ready for Test T2?

Section 4 An Overview of the System

Section 5 Sample Test for Practice

Section 6 Additional Test Questions for Practice

Section 7 Appendices

Preface

This book is just one of a comprehensive series designed to prepare technicians to take and pass every ASE test. Delmar's series covers all of the Automotive tests A1 through A8 as well as Advanced Engine Performance L1 and Parts Specialist P2. The series also covers the five Collision Repair tests and the eight Medium/Heavy Duty truck test.

Before any book in this series was written, Delmar staff met with and surveyed technicians and shop owners who have taken ASE tests and have used other preparatory materials. We found that they wanted, first and foremost, *lots* of practice tests and questions. Each book in our series contains a general knowledge pretest, a sample test, and additional practice questions. You will be hard-pressed to find a test prep book with more questions for you to practice with. We have worked hard to ensure that these questions match the ASE style in types of questions, quantities, and level of difficulty.

Technicians also told us that they wanted to understand the ASE test and to have practical information about what they should expect. We have provided that as well, including a history of ASE and a section devoted to helping the technician "Take and Pass Every ASE Test" with case studies, test-taking strategies, and test formats.

Finally, techs wanted refresher information and reference. Each of our books includes an overview section that is referenced to the task list. The complete task lists for each test appear in each book for the user's reference. There is also a complete glossary of terms for each booklet.

So whether you're looking for a sample test and a few extra questions to practice with or a complete introduction to ASE testing, with support for preparing thoroughly, this book series is an excellent answer.

We hope you benefit from this book and that you pass every ASE test you take!

Your comments, both positive and negative, are certainly encouraged! Please contact us at:

Automotive Editor
Delmar Publishers
3 Columbia Circle
Box 15015
Albany, NY 12212-5015

The History of ASE

History

Originally known as The National Institute for Automotive Service Excellence (NIASE), today's ASE was founded in 1972 as a nonprofit, independent entity dedicated to improving the quality of automotive service and repair through the voluntary testing and certification of automotive technicians. Until that time, consumers had no way of distinguishing between competent and incompetent automotive technicians. In the mid-1960s and early 1970s, efforts were made by several automotive industry affiliated associations to respond to this need. Though the associations were nonprofit, many regarded certification test fees merely as a means of raising additional operating capital. Also, some associations, having a vested interest, produced test scores heavily weighted in the favor of its members.

NIASE

From these efforts a new independent, nonprofit association, the National Institute for Automotive Service Excellence (NIASE), was established much to the credit of two educators, George R. Kinsler, Director of Program Development for the Wisconsin Board of Vocational and Adult Education in Madison, WI, and Myron H. Appel, Division Chairman at Cypress College in Cypress, CA.

Early efforts were to encourage voluntary certification in four general areas:

TEST AREA	TITLES
I. Engine	Engines, Engine Tune-Up, Block Assembly, Cooling and Lube Systems, Induction, Ignition, and Exhaust
II. Transmission	Manual Transmissions, Drive Line and Rear Axles, and Automatic Transmissions
III. Brakes and Suspension	Brakes, Steering, Suspension, and Wheels
IV. Electrical/Air Conditioning	Body/Chassis, Electrical Systems, Heating, and Air Conditioning

In early NIASE tests, Mechanic A, Mechanic B type questions were used. Over the years the trend has not changed, but in mid-1984 the term was changed to Technician A, Technician B to better emphasize sophistication of the skills needed to perform successfully in the modern motor vehicle industry. In certain tests the term used is Estimator A/B, Painter A/B, or Parts Specialist A/B. At about that same time, the logo was changed from "The Gear" to "The Blue Seal," and the organization adopted the acronym ASE for Automotive Service Excellence.

Since those early beginnings, several other related trades have been added. ASE now administers a comprehensive series of certification exams for automotive and light truck repair technicians, medium and heavy truck repair technicians, alternate fuels technicians, engine machinists, collision repair technicians, school bus repair technicians, and parts specialists.

The Series and Individual Tests

- Automotive and Light Truck Technician; consisting of: Engine Repair—Automatic Transmission/Transaxle—Manual Drive Train and Axles—Suspension and Steering—Brakes—Electrical/Electronic Systems—Heating and Air Conditioning—Engine Performance
- Medium and Heavy Truck Technician; consisting of: Gasoline Engines—Diesel Engines—Drive Train—Brakes—Suspension and Steering—Electrical/Electronic Systems—HVAC—Preventive Maintenance Inspection
- Alternate Fuels Technician; consisting of: Compressed Natural Gas Light Vehicles
- Advanced Series; consisting of: Automobile Advanced Engine Performance and Advanced Diesel Engine Electronic Diesel Engine Specialty
- Collision Repair Technician; consisting of: Painting and Refinishing—Non-Structural Analysis and Damage Repair—Structural Analysis and Damage Repair—Mechanical and Electrical Components—Damage Analysis and Estimating
- Engine Machinist Technician; consisting of: Cylinder Head Specialist—Cylinder Block Specialist—Assembly Specialist
- School Bus Repair Technician; consisting of: Body Systems and Special Equipment—Drive Train—Brakes—Suspension and Steering—Electrical/Electronic Systems—Heating and Air Conditioning
- Parts Specialist; consisting of: Automobile Parts Specialist—Medium/Heavy Truck Parts Specialist

A Brief Chronology

1970–1971 Original questions were prepared by a group of forty auto mechanics teachers from public secondary schools, technical institutes, community colleges, and private vocational schools. These questions were then professionally edited by testing specialists at Educational Testing Service (ETS) at Princeton, New Jersey, and thoroughly reviewed by training specialists associated with domestic and import automotive companies.

1971 July: About eight hundred mechanics tried out the original test questions at experimental test administrations.

1972 November and December: Initial NIASE tests administered at 163 test centers. The original automotive test series consisted of four tests containing eighty questions each. Three hours were allotted for each test. Those who passed all four tests were designated Certified General Auto Mechanic (GAM).

1973 April and May: Test 4 was increased to 120 questions. Time was extended to four hours for this test. There were now 182 test centers. Shoulder patch insignias were made available.

	November: Automotive series expanded to five tests. Heavy-Duty Truck series of six tests introduced.
1974	November: Automatic Transmission (Light Repair) test modified. Name changed to Automatic Transmission.
1975	May: Collision Repair series of two tests is introduced.
1978	May: Automotive recertification testing is introduced.
1979	May: Heavy-Duty Truck recertification testing is introduced.
1980	May: Collision Repair recertification testing is introduced.
1982	May: Test administration providers switched from Educational Testing Service (ETS) to American College Testing (ACT). Name of Automobile Engine Tune-Up test changed to Engine Performance test.
1984	May: New logo was introduced. ASE's "The Blue Seal" replaced NIASE's "The Gear." All reference to Mechanic A, Mechanic B was changed to Technician A, Technician B.
1990	November: The first of the Engine Machinist test series was introduced.
1991	May: The second test of the Engine Machinist test series was introduced. November: The third and final Engine Machinist test was introduced.
1992	May: Name of Heavy-Duty Truck Test series changed to Medium/Heavy Truck test series.
1993	May: Automotive Parts Specialist test introduced. Collision Repair expanded to six tests. Light Vehicle Compressed Natural Gas test introduced. Limited testing begins in English-speaking provinces of Canada.
1994	May: Advanced Engine Performance Specialist test introduced.
1996	May: First three tests for School Bus Technician test series introduced. November: A Collision Repair test is added.
1997	May: A Medium/Heavy Truck test is added.
1998	May: A diesel advanced engine test is introduced: Electronic Diesel Engine Diagnosis Specialist. A test is added to the School Bus test series.

By the Numbers

Following are the approximate number of ASE technicians currently certified by category. The numbers may vary from time to time but are reasonably accurate for any given period. More accurate data may be obtained from ASE, which provides updates twice each year, in May and November after the Spring and Fall test series.

There are more than 338,000 Automotive Technicians with over 87,000 at Master Technician (MA) status. There are 47,000 Truck Technicians with over 19,000 at Master Technician (MT) status. There are 46,000 Collision Repair/Refinish Technicians with 7,300 at Master Technician (MB) status. There are 1,200 Estimators. There are 6,700 Engine Machinists with over 2,800 at Master Machinist Technician (MM) status. There are also 28,500 Automobile Advanced Engine Performance Technicians and over 2,700 School Bus Technicians for a combined total of more than 403,000 Repair Technicians. To this number, add over 22,000 Automobile Parts Specialists, and over 2,000 Truck Parts Specialists for a combined total of over 24,000 parts specialists.

There are over 6,400 ASE Technicians holding both Master Automotive Technician and Master Truck Technician status, of which 350 also hold Master Body Repair status. Almost 200 of these Master Technicians also hold Master Machinist status and five Technicians are certified in all ASE specialty areas.

Almost half of ASE certified technicians work in new vehicle dealerships (45.3 percent). The next greatest number work in independent garages with 19.8 percent. Next is tire dealerships with 9 percent, service stations at 6.3 percent, fleet shops at 5.7 percent, franchised volume retailers at 5.4 percent, paint and body shops at 4.3 percent, and specialty shops at 3.9 percent.

Of over 400,000 automotive technicians on ASE's certification rosters, almost 2,000 are female. The number of female technicians is increasing at a rate of about 20 percent each year. Women's increasing interest in the automotive industry is further evidenced by the fact that, according to the National Automobile Dealers Association (NADA), they influence 80 percent of the decisions of the purchase of a new automobile and represent 50 percent of all new car purchasers. Also, it is interesting to note that 65 percent of all repair and maintenance service customers are female.

The typical ASE certified technician is 36.5 years of age, is computer literate, deciphers a half-million pages of technical manuals, spends one hundred hours per year in training, holds four ASE certificates, and spends about $27,000 for tools and equipment. Twenty-seven percent of today's skilled ASE certified technicians attended college, many having earned an Associate of Science degree in Automotive Technology.

ASE

ASE's mission is to improve the quality of vehicle repair and service in the United States through the testing and certification of automotive repair technicians. Prospective candidates register for and take one or more of ASE's thirty-three exams. The tests are grouped into specialties for automobile, medium/heavy truck, school bus, and collision repair technicians as well as engine machinists, alternate fuels technicians, and parts specialists.

Upon passing at least one exam and providing proof of two years of related work experience, the technician becomes ASE certified. A technician who passes a series of exams earns ASE Master Technician status. An automobile technician, for example, must pass eight exams for this recognition.

The tests, conducted twice a year at over seven hundred locations around the country, are administered by American College Testing (ACT). They stress real-world diagnostic and repair problems. Though a good knowledge of theory is helpful to the technician in answering many of the questions, there are no questions specifically on theory. Certification is valid for five years. To retain certification, the technician must be retested to renew his or her certificate.

The automotive consumer benefits because ASE certification is a valuable yardstick by which to measure the knowledge and skills of individual technicians, as well as their commitment to their chosen profession. It is also a tribute to the repair facility employing ASE certified technicians. ASE certified technicians are permitted to wear blue and white ASE shoulder insignia, referred to as the "Blue Seal of Excellence," and carry credentials listing their areas of expertise. Often employers display their technicians' credentials in the customer waiting area. Customers look for facilities that display ASE's Blue Seal of Excellence logo on outdoor signs, in the customer waiting area, in the telephone book (Yellow Pages), and in newspaper advertisements.

The tests stress repair knowledge and skill. All test takers are issued a score report. In order to earn ASE certification, a technician must pass one or more of the exams and present proof of two years of relevant hands-on work experience. ASE certifications are valid for five years, after which time technicians must retest in order to keep up with changing technology and to remain in the ASE program. A nominal registration and test fee is charged.

To become part of the team that wears ASE's Blue Seal of Excellence®, please contact:

National Institute for Automotive Service Excellence
13505 Dulles Technology Drive
Herndon, VA 20171-3421

2 Take and Pass Every ASE Test

ASE Testing

Participating in an Automotive Service Excellence (ASE) voluntary certification program gives you a chance to show your customers that you have the "know-how" needed to work on today's modern vehicles. The ASE certification tests allow you to compare your skills and knowledge to the automotive service industry's standards for each specialty area.

If you are the "average" automotive technician taking this test, you are in your mid-thirties and have not attended school for about fifteen years. That means you probably have not taken a test in many years. Some of you, on the other hand, have attended college or taken postsecondary education courses and may be more familiar with taking tests and with test-taking strategies. There is, however, a difference in the ASE test you are preparing to take and the educational tests you may be accustomed to.

Who Writes the Questions?

The questions on an educational test are generally written, administered, and graded by an educator who may have little or no practical hands-on experience in the test area. The questions on all ASE tests are written by service industry experts familiar with all aspects of the subject area. ASE questions are entirely job-related and designed to test the skills that you need to know on the job.

The questions originate in an ASE "item-writing" workshop where service representatives from domestic and import automobile manufacturers, parts and equipment manufacturers, and vocational educators meet in a workshop setting to share their ideas and translate them into test questions. Each test question written by these experts is reviewed by all of the members of the group. The questions deal with the practical problems of diagnosis and repair that are experienced by technicians in their day-to-day hands-on work experiences.

All of the questions are pretested and quality-checked in a nonscoring section of tests by a national sample of certifying technicians. The questions that meet ASE's high standards of accuracy and quality are then included in the scoring sections of future tests. Those questions that do not pass ASE's stringent tests are sent back to the workshop or are discarded. ASE's tests are monitored by an independent proctor and are administered and machine-scored by an independent provider, American College Testing (ACT). All ASE tests have a three-year revision cycle.

Testing

If you think about it, we are actually tested on about everything we do. As infants, we were tested to see when we could turn over and crawl, later when we could walk or talk. As adolescents, we were tested to determine how well we learned the material presented in school and in how we demonstrated our accomplishments on the athletic field. As working adults, we are tested by our supervisors on how well we have completed an assignment or project. As nonworking adults, we are tested by our family on everyday activities, such as housekeeping or preparing a meal. Testing, then, is one of those facts of life that begins in the cradle and follows us to the grave.

Testing is an important fact of life that helps us to determine how well we have learned our trade. Also, tests often help us to determine what opportunities will be available to us in the future. To become ASE certified, we are required to take a test in every subject in which we wish to be recognized.

Be Test-Wise

In spite of the widespread use of tests, most technicians are not very test-wise. An ability to take tests and score well is a skill that must be acquired. Without this knowledge, the most intelligent and prepared technician may not do well on a test.

We will discuss some of the basic procedures necessary to follow in order to become a test-wise technician. Assume, if you will, that you have done the necessary study and preparation to score well on the ASE test.

Different approaches should be used for taking different types of tests. The different basic types of tests include: essay, objective, multiple-choice, fill-in-the-blank, true-false, problem solving, and open book. All ASE tests are of the four-part multiple-choice type.

Before discussing the multiple-choice type test questions, however, there are a few basic principles that should be followed before taking any test.

Before the Test

Do not arrive late. Always arrive well before your test is scheduled to begin. Allow ample time for the unexpected, such as traffic problems, so you will arrive on time and avoid the unnecessary anxiety of being late.

Always be certain to have plenty of supplies with you. For an ASE test, three or four sharpened soft lead (#2) pencils, a pocket pencil sharpener, erasers, and a watch are all that are required.

Do not listen to pretest chatter. When you arrive early, you may hear other technicians testing each other on various topics or making their best guess as to the probable test questions. At this time, it is too late to add to your knowledge. Also the rhetoric may only confuse you. If you find it bothersome, take a walk outside the test room to relax and loosen up.

Read and listen to all instructions. It is important to read and listen to the instructions. Make certain that you know what is expected of you. Listen carefully to verbal instructions and pay particular attention to any written instructions on the test paper. Do not dive into answering questions only to find out that you have answered the wrong question by not following instructions carefully. It is difficult to make a high score on a test if you answer the wrong questions.

These basic principles have been violated in most every test ever given. Try to remember them. They are essential for success.

Objective Tests

A test is called an objective test if the same standards and conditions apply to everyone taking the test and there is only one correct answer to each question. Objective tests primarily measure your ability to recall information. A well-designed objective test can also test your ability to understand, analyze, interpret, and apply your knowledge. Objective tests include true-false, multiple-choice, fill-in-the-blank, and matching questions.

Objective questions, not generally encountered in a classroom setting, are frequently used in standardized examinations. Objective tests are easy to grade and also reduce the amount of paperwork necessary to administer. The objective tests are used in entry-level programs or when very large numbers are being tested. ASE's tests consist exclusively of four-part multiple-choice objective questions in all of their tests.

Taking an Objective Test

The principles of taking an objective test are somewhat different from those used in other types of tests. You should first quickly look over the test to determine the number of questions, but do not try to read through all of the questions. In an ASE test, there are usually between forty and eighty questions, depending on the subject matter. Read through each question before marking your answer. Answer the questions in the order they appear on the test. Leave the questions blank that you are not sure of and move on to the next question. You can return to those unanswered questions after you have finished the others. They may be easier to answer at a later time after your mind has had additional time to consider them on a subconscious level. In addition, you might find information in other questions that will help you to answer some of them.

Do not be obsessed by the apparent pattern of responses. For example, do not be influenced by a pattern like **d**, **c**, **b**, **a**, **d**, **c**, **b**, **a** on an ASE test.

There is also a lot of folk wisdom about taking objective tests. For example, there are those who would advise you to avoid response options that use certain words such as *all, none, always, never, must,* and *only,* to name a few. This, they claim, is because nothing in life is exclusive. They would advise you to choose response options that use words that allow for some exception, such as *sometimes, frequently, rarely, often, usually, seldom,* and *normally*. They would also advise you to avoid the first and last option (A and D) because test writers, they feel, are more comfortable if they put the correct answer in the middle (B and C) of the choices. Another recommendation often offered is to select the option that is either shorter or longer than the other three choices because it is more likely to be correct. Some would advise you to never change an answer since your first intuition is usually correct.

Although there may be a grain of truth in this folk wisdom, ASE test writers try to avoid them and so should you. There are just as many **A** answers as there are **B** answers, just as many **D** answers as **C** answers. As a matter of fact, ASE tries to balance the answers at about 25 percent per choice **A**, **B**, **C**, and **D**. There is no intention to use "tricky" words, such as outlined above. Put no credence in the opposing words "sometimes" and "never," for example. When used in an ASE type question, one or both may be correct; one or both may be incorrect.

There are some special principles to observe on multiple-choice tests. These tests are sometimes challenging because there are often several choices that may seem possible, and it may be difficult to decide on the correct choice. The best strategy, in this case, is to first determine the correct answer before looking at the options. If you see the answer you decided on, you should still examine the options to make sure that none seem more correct than yours. If you do not know or are not sure of the answer, read each option very carefully and try to eliminate those options that you know to be wrong. That way, you can often arrive at the correct choice through a process of elimination.

If you have gone through all of the test and you still do not know the answer to some of the questions, then guess. Yes, guess. You then have at least a 25 percent chance of being correct. If you leave the question blank, you have no chance. In ASE tests, there is no penalty for being wrong. As the late President Franklin D. Roosevelt once advised a group of students, "It is common sense to take a method and try it. If it fails, admit it frankly and try another. But above all, try something."

During the Test

Mark your bubble sheet clearly and accurately. One of the biggest problems an adult faces in test-taking, it seems, is in placing an answer in the correct spot on a bubble sheet. Make certain that you mark your answer for, say, question 21, in the space on the bubble sheet designated for the answer for question 21. A correct response in the wrong bubble will probably be wrong. Remember, the answer sheet is machine scored and can only "read" what you have bubbled in. Also, do not bubble in two answers for the same question. For example, if you feel the answer to a particular question is **A** but think it may be **C,** do not bubble in both choices. Even if either **A** or **C** is correct, a double answer will score as an incorrect answer. It's better to take a chance with your best guess.

Review Your Answers

If you finish answering all of the questions on a test ahead of time, go back and review the answers of those questions that you were not sure of. You can often catch careless errors by using the remaining time to review your answers.

Don't Be Distracted

At practically every test, some technicians will invariably finish ahead of time and turn their papers in long before the final call. Do not let them distract or intimidate you. Either they knew too little and could not finish the test, or they were very self-confident and thought they knew it all. Perhaps they were trying to impress the proctor or other technicians about how much they know. Often you may hear them later talking about the information they knew all the while but forgot to respond on their answer sheet.

Use Your Time Wisely

It is not wise to use less than the total amount of time that you are allotted for a test. If there are any doubts, take the time for review. Any product can usually be made better with some additional effort. A test is no exception. It is not necessary to turn in your test paper until you are told to do so.

Don't Cheat

Some technicians may try to use a "crib sheet" during a test. Others may attempt to read answers from another technician's paper. If you do that, you are unquestionably assuming that someone else has a correct answer. You probably know as much, maybe more, than anyone else in the test room. Trust yourself. If you're still not convinced, think of the consequences of being caught. Cheating is foolish. If you are caught, you have failed the test.

Be Confident

The first and foremost principle in taking a test is that you need to know what you are doing, to be test-wise. It will now be presumed that you are a test-wise technician and are now ready for some of the more obscure aspects of test-taking.

An ASE-style test requires that you use the information and knowledge at your command to solve a problem. This generally requires a combination of information similar to the way you approach problems in the real world. Most problems, it seems, typically do not fall into neat textbook cases. New problems are often difficult to handle, whether they are encountered inside or outside the test room.

An ASE test also requires that you apply methods taught in class as well as those learned on the job to solve problems. These methods are akin to a well-equipped tool box in the hands of a skilled technician. You have to know what tools to use in a particular situation, and you must also know how to use them. In an ASE test, you will need to be able to demonstrate that you are familiar with and know how to use the tools.

You should begin a test with a completely open mind. At times, however, you may have to move out of your normal way of thinking and be creative to arrive at a correct answer. If you have diligently studied for at least one week prior to the test, you have bombarded your mind with a wide assortment of information. Your mind will be working with this information on a subconscious level, exploring the interrelationships among various facts, principles, and ideas. This prior preparation should put you in a creative mood for the test.

In order to reach your full potential, you should begin a test with the proper mental attitude and a high degree of self-confidence. You should think of a test as an opportunity to document how much you know about the various tasks in your chosen profession. If you have been diligently studying the subject matter, you will be able to take your test in serenity because your mind will be well organized. If you are confident, you are more likely to do well because you have the proper mental attitude. If, on the other hand, your confidence is low, you are bound to do poorly. It is a self-fulfilling prophecy.

Perhaps you have heard athletic coaches talk about the importance of confidence when competing in sports. Mental confidence helps an athlete to perform at the highest level and gain an advantage over competitors. Taking a test is much like an athletic event. You are competing against yourself, in a certain sense, because you will be trying to approach perfection in determining your answers. As in any competition, you should aim your sights high and be confident that you can reach the apex.

Anxiety and Fear

Many technicians experience anxiety and fear at the very thought of taking a test. Many worry, become nervous, and even become ill at test time because of the fear of failure. Many often worry about the criticism and ridicule that may come from their employer, relatives, and peers. Some worry about taking a test because they feel that the stakes are very high. Those who spent a great amount of time studying may feel they must get a high grade to justify their efforts. The thought of not doing well can result in unnecessary worry. They become so worried, in fact, that their reasoning and thinking ability is impaired, actually bringing about the problem they wanted to avoid.

The fear of failure should not be confused with the desire for success. It is natural to become "psyched-up" for a test in contemplation of what is to come. A little emotion can provide a healthy flow of adrenaline to peak your senses and hone your mental ability. This improves your performance on the test and is a very different reaction from fear.

Most technician's fears and insecurities experienced before a test are due to a lack of self-confidence. Those who have not scored well on previous tests or have no confidence in their preparation are those most likely to fail. Be confident that you will do well on your test and your fears should vanish. You will know that you have done everything possible to realize your potential.

Getting Rid of Fear

If you have previously experienced fear of taking a test, it may be difficult to change your attitude immediately. It may be easier to cope with fear if you have a better understanding of what the test is about. A test is merely an assessment of how much the technician knows about a particular task area. Tests, then, are much less threatening when thought of in this manner. This does not mean, however, that you should lower your self-esteem simply because you performed poorly on a test.

You can consider the test essentially as a learning device, providing you with valuable information to evaluate your performance and knowledge. Recognize that no one is perfect. All humans make mistakes. The idea, then, is to make mistakes before the test, learn from them, and avoid repeating them on the test. Fortunately, this is not as difficult as it seems. Practical questions in this study guide include the correct answers to consider if you have made mistakes on the practice test. You should learn where you went wrong so you will not repeat them in the ASE test. If you learn from your mistakes, the stage is set for future growth.

If you understood everything presented up until now, you have the knowledge to become a test-wise technician, but more is required. To be a test-wise technician, you not only have to practice these principles, you have to diligently study in your task area.

Effective Study

The fundamental and vital requirement to induce effective study is a genuine and intense desire to achieve. This is more basic than any rule or technique that will be given here. The key requirement, then, is a driving motivation to learn and to achieve.

If you wish to study effectively, first develop a desire to master your studies and sincerely believe that you will master them. Everything else is secondary to such a desire.

First, build up definite ambitions and ideals toward which your studies can lead. Picture the satisfaction of success. The attitude of the technician may be transformed from merely getting by to an earnest and energetic effort. The best direct stimulus to change may involve nothing more than the deliberate planning of your time. Plan time to study.

Another drive that creates positive study is an interest in the subject studied. As an automotive technician, you can develop an interest in studying particular subjects if you follow these four rules:

1. Acquire information from a variety of sources. The greater your interest in a subject, the easier it is to learn about it. Visit your local library and seek books on the subject you are studying. When you find something new or of interest, make inexpensive photocopies for future study.
2. Merge new information with your previous knowledge. Discover the relationship of new facts to old known facts. Modern developments in automotive technology take on new interest when they are seen in relation to present knowledge.
3. Make new information personal. Relate the new information to matters that are of concern to you. The information you are now reading, for example, has interest to you as you think about how it can help.
4. Use your new knowledge. Raise questions about the points made by the book. Try to anticipate what the next steps and conclusions will be. Discuss this new knowledge, particularly the difficult and questionable points, with your peers.

You will find that when you study with eager interest, you will discover it is no longer work. It is pleasure and you will be fascinated in what you study. Studying can be like reading a novel or seeing a movie that overcomes distractions and requires no effort or willpower. You will discover that the positive relationship between interest and effort works both ways. Even though you perhaps began your studies with little or no interest, simply staying with it helped you to develop an interest in your studies.

Obviously, certain subject matter studies are bound to be of little or no interest, particularly in the beginning. Parts of certain studies may continue to be uninteresting. An honest effort to master those subjects, however, nearly always brings about some level of interest. If you appreciate the necessity and reward of effective studying, you will rarely be disappointed. Here are a few important hints for gaining the determination that is essential to carrying good conclusions into actual practice.

Make Study Definite

Decide what is to be studied and when it is to be studied. If the unit is discouragingly long, break it into two or more parts. Determine exactly what is involved in the first part and learn that. Only then should you proceed to the next part. Stick to a schedule.

The Urge to Learn

Make clear to yourself the relation of your present knowledge to your study materials. Determine the relevance with regard to your long-range goals and ambitions.

Turn your attention away from real or imagined difficulties as well as other things that you would rather be doing. Some major distractions are thoughts of other duties and of disturbing problems. These distractions can usually be put aside, simply shunted off by listing them in a notebook. Most technicians have found that by writing interfering thoughts down, their minds are freed from annoying tensions.

Adopt the most reasonable solution you can find or seek objective help from someone else for personal problems. Personal problems and worry are often causes of ineffective study. Sometimes there are no satisfactory solutions. Some manage to avoid the problems or to meet them without great worry. For those who may wish to find better ways of meeting their personal problems, the following suggestions are offered:

1. Determine as objectively and as definitely as possible where the problem lies. What changes are needed to remove the problem, and which changes, if any, can be made? Sometimes it is wiser to alter your goals than external conditions. If there is no perfect solution, explore the others. Some solutions may be better than others.
2. Seek an understanding confidant who may be able to help analyze and meet your problems. Very often, talking over your problems with someone in whom you have confidence and trust will help you to arrive at a solution.
3. Do not betray yourself by trying to evade the problem or by pretending that it has been solved. If social problem distractions prevent you from studying or doing satisfactory work, it is better to admit this to yourself. You can then decide what can be done about it.

Once you are free of interferences and irritations, it is much easier to stay focused on your studies.

Concentrate

To study effectively, you must concentrate. Your ability to concentrate is governed, to a great extent, by your surroundings as well as your physical condition. When absorbed in study, you must be oblivious to everything else around you. As you learn to concentrate and study, you must also learn to overcome all distractions. There are three kinds of distractions you may face:

1. Distractions in the surrounding area, such as motion, noise, and the glare of lights. The sun shining through a window on your study area, for example, can be very distracting.

   Some technicians find that, for effective study, it is necessary to eliminate visual distractions as well as noises. Others find that they are able to tolerate moderate levels of auditory or visual distraction.

   Make sure your study area is properly lighted and ventilated. The lighting should be adequate but should not shine directly into your eyes or be visible out of the corner of your eye. Also, try to avoid a reflection of the lighting on the pages of your book.

   Whether heated or cooled, the environment should be at a comfortable level. For most, this means a temperature of 78°F–80°F (25.6°C–26.7°C) with a relative humidity of 45 to 50 percent.

2. Distractions arising from your body, such as a headache, fatigue, and hunger. Be in good physical condition. Eat wholesome meals at regular times. Try to eat with your family or friends whenever possible. Mealtime should be your recreational period. Do not eat a heavy meal for lunch, and do not resume studies immediately after eating lunch. Just after lunch, try to get some regular exercise, relaxation, and recreation. A little exercise on a regular basis is much more valuable than a lot of exercise only on occasion.
3. Distractions of irrelevant ideas, such as how to repair the garden gate, when you are studying for an automotive-related test.

The problems associated with study are no small matter. These problems of distractions are generally best dealt with by a process of elimination. A few important rules for eliminating distractions follow.

Get Sufficient Sleep

You must get plenty of rest even if it means dropping certain outside activities. Avoid cutting in on your sleep time; you will be rewarded in the long run. If you experience difficulty going to sleep, do something to take your mind off your work and try to relax before going to bed. Some suggestions that may help include a little reading, a warm bath, a short walk, a conversation with a friend, or writing that overdue letter to a distant relative. If sleeplessness is an ongoing problem, consult a physician. Do not try any of the sleep remedies on the market, particularly if you are on medication, without approval of your physician.

If you still have difficulty studying, a final rule may help. Sit down in a favorable place for studying, open your books, and take out your pencil and paper. In a word, go through the motions.

Arrange Your Area

Arrange your chair and work area. To avoid strain and fatigue, whenever possible, shift your position occasionally. Try to be comfortable; however, avoid being too comfortable. It is nearly impossible to study rigorously when settled back in a large easy chair or reclining leisurely on a sofa.

When studying, it is essential to have a plan of action, a time to work, a time to study, and a time for pleasure. If you schedule your day and adhere to the schedule, you will eliminate most of your efforts and worries. A plan that is followed, then, soon becomes the easy and natural routine of the day. Most technicians find it useful to have a definite place and time to study. A particular table and chair should always be used for study and intellectual work. This place will then come to mean study. To be seated in that particular location at a regularly scheduled time will automatically lead you to assume a readiness for study.

Don't Daydream

Daydreaming or mind-wandering is an enemy of effective study. Daydreaming is frequently due to an inadequate understanding of words. Use the Glossary or a dictionary to look up the troublesome word. Another frequent cause of daydreaming is a deficient background in the present subject matter. When this is the problem, go back and review the subject matter to obtain the necessary foundation. Just one hour of concentrated study is equivalent to ten hours with frequent lapses of daydreaming. Be on guard against mind-wandering, and pull yourself back into focus on every occasion.

Study Regularly

A system of regularity in study is believed by many scholars to be the secret of success. The daily time schedule must, however, be determined on an individual basis. You must decide how many hours of each day you can devote to your studies. Few technicians really are aware of where their leisure time is spent. An accurate account of how your days are presently being spent is an important first step toward creating an effective daily schedule.

Weekly Schedule							
	Sun	Mon	Tues	Wed	Thu	Fri	Sat
6:00							
6:30							
7:00							
7:30							
8:00							
8:30							
9:00							
9:30							
10:00							
10:30							
11:00							
11:30							
NOON							
12:30							
1:00							
1:30							
2:00							
2:30							
3:00							
3:30							
4:00							
4:30							
5:00							
5:30							
6:00							
6:30							
7:00							
7:30							
8:00							
8:30							
9:00							
9:30							
10:00							
10:30							
11:00							
11:30							

The convenient form is for keeping an hourly record of your week's activities. If you fill in the schedule each evening before bedtime, you will soon gain some interesting and useful facts about yourself and your use of your time. If you think over the causes of wasted time, you can determine how you might better spend your time. A practical schedule can be set up by using the following steps.

1. Mark your fixed commitments, such as work, on your schedule. Be sure to include classes and clubs. Do you have sufficient time left? You can arrive at an estimate of the time you need for studying by counting the hours used during the present week. An often-used formula, if you are taking classes, is to multiply the number of hours you spend in class by two. This provides time for class studies. This is then added to your work hours. Do not forget time allocation for travel.
2. Fill in your schedule for meals and studying. Use as much time as you have available during the normal workday hours. Do not plan, for example, to do all of your studying between 11:00 P.M. and 1:00 A.M. Try to select a time for study that you can use every day without interruption. You may have to use two or perhaps three different study periods during the day.
3. List the things you need to do within a time period. A one-week time frame seems to work well for most technicians. The question you may ask yourself is: "What do I need to do to be able to walk into the test next week, or next month, prepared to pass?"
4. Break down each task into smaller tasks. The amount of time given to each area must also be settled. In what order will you tackle your schedule? It is best to plan the approximate time for your assignments and the order in which you will do them. In this way, you can avoid the difficulties of not knowing what to do first and of worrying about the other things you should be doing.
5. List your tasks in the empty spaces on your schedule. Keep some free time unscheduled so you can deal with any unexpected events, such as a dental appointment. You will then have a tentative schedule for the following week. It should be flexible enough to allow some units to be rearranged if necessary. Your schedule should allow time off from your studies. Some use the promise of a planned recreational period as a reward for motivating faithfulness to a schedule. You will more likely lose control of your schedule if it is packed too tightly.

Keep a Record

Keep a record of what you actually do. Use the knowledge you gain by keeping a record of what you are actually doing so you can create or modify a schedule for the following week. Be sure to give yourself credit for movement toward your goals and objectives. If you find that you cannot study productively at a particular hour, modify your schedule so as to correct that problem.

Scoring the ASE Test

You can gain a better perspective about tests if you know and understand how they are scored. ASE's tests are scored by American College Testing (ACT), a nonpartial, nonbiased organization having no vested interest in ASE or in the automotive industry. Each question carries the same weight as any other question. For example, if there are fifty questions, each is worth 2 percent of the total score. The passing grade is 70 percent. That means you must correctly answer thirty-five of the fifty questions to pass the test.

Understand the Test Results

The test results can tell you:

- where your knowledge equals or exceeds that needed for competent performance, or
- where you might need more preparation.

The test results *cannot* tell you:

- how you compare with other technicians, or
- how many questions you answered correctly.

Your ASE test score report will show the number of correct answers you got in each of the content areas. These numbers provide information about your performance in each area of the test. However, because there may be a different number of questions in each area of the test, a high percentage of correct answers in an area with few questions may not offset a low percentage in an area with many questions.

It may be noted that one does not "fail" an ASE test. The technician that does not pass is simply told "More Preparation Needed." Though large differences in percentages may indicate problem areas, it is important to consider how many questions were asked in each area. Since each test evaluates all phases of the work involved in a service specialty, you should be prepared in each area. A low score in one area could keep you from passing an entire test.

Note that a typical test will contain the number of questions indicated above each content area's description. For example:

Diesel Engines (Test T2)

Content Area	Questions	Percent of Test
A. General Engine Diagnosis	16	20%
B. Cylinder Head and Valve Train Diagnosis and Repair	8	10%
C. Engine Block Diagnosis and Repair	8	10%
D. Lubrication and Cooling Systems Diagnosis and Repair	11	14%
E. Air Induction and Exhaust Systems Diagnosis and Repair	11	14%
F. Fuel System Diagnosis and Repair	20	25%
1. Mechanical Components (10)		
2. Electronic Components (10)		
G. Starting System Diagnosis and Repair	4	5%
H. Engine Brakes	2	2%
Total	*80	100%

***Note:** The test could contain up to ten additional questions that are included for statistical research purposes only. Your answers to these questions will not affect your score, but since you do not know which ones they are, you should answer all questions in the test. The five-year Recertification Test will cover the same content areas as those listed above. However, the number of questions in each content area of the Recertification Test will be reduced by about one-half.*

"Average"

There is no such thing as average. You cannot determine your overall test score by adding the percentages given for each task area and dividing by the number of areas. It doesn't work that way because there generally are not the same number of questions in each task area. A task area with twenty questions, for example, counts more toward your total score than a task area with ten questions.

So, How Did You Do?

Your test report should give you a good picture of your results and a better understanding of your task areas of strength and weakness.

If you fail to pass the test, you may take it again at any time it is scheduled to be administered. You are the only one who will receive your test score. Test scores will not be given over the telephone by ASE nor will they be released to anyone without your written permission.

3 Are You Sure You're Ready for Test T2?

Pretest

The purpose of this pretest is to determine the amount of review that you may require prior to taking the ASE medium/heavy truck test: Diesel Engines (Test T2). If you answer all of the pretest questions correctly, complete the sample test in section 5 along with the additional test questions in section 6.

If two or more of your answers to the pretest questions are wrong, study section 4: An Overview of the System before continuing with the sample test and additional test questions.

The pretest answers and explanations are located at the end of the pretest.

1. To inspect a cylinder head for cracks, all of the following may be necessary **EXCEPT:**
   A. use a putty knife to remove excess gasket material from the head and inspect for visible cracks between cylinders and ports.
   B. check the torque of head bolts.
   C. use magnetic or dye crack detection to find small cracks that are not otherwise visible.
   D. remove excess carbon buildup by sanding.
2. Excessive surges in the coolant level could be caused by all of the following **EXCEPT:**
   A. a loose fan belt.
   B. a faulty fuel injector.
   C. a restricted radiator.
   D. a blown head gasket.
3. A diesel engine is idling lower than specifications. Technician A says this could be caused by worn governor parts. Technician B says that a clogged fuel filter could be the cause. Who is right?
   A. A only
   B. B only
   C. Both A and B
   D. Neither A nor B
4. Technician A says that excessive smoke from a diesel engine indicates the engine needs to be overhauled. Technician B says that by discussing the engine operation with the operator you may find that an engine overhaul is not needed. Who is right?
   A. A only
   B. B only
   C. Both A and B
   D. Neither A nor B

5. Which one of the following diesel engine types uses an O-ring to prevent coolant from leaking into the crankcase?
   A. Dry sleeve
   B. Wet sleeve
   C. Parent bore
   D. Throw away
6. The valves of a diesel engine must be replaced if:
   A. there is pitting on the valve face.
   B. the engine is undergoing an overhaul.
   C. the seats are to be resurfaced.
   D. they show signs of cracking or warpage.
7. What would most likely be the reason for replacing a sleeve in a diesel engine?
   A. Normal wear
   B. Taper
   C. Damage
   D. Out-of-roundness
8. What is the most likely cause of bearing failure?
   A. Overloading
   B. Corrosion
   C. Dirt
   D. Misalignment

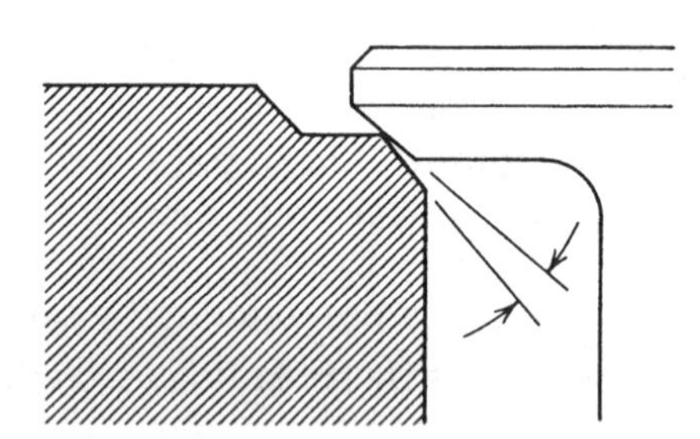

Intake Valve And Valve Seat

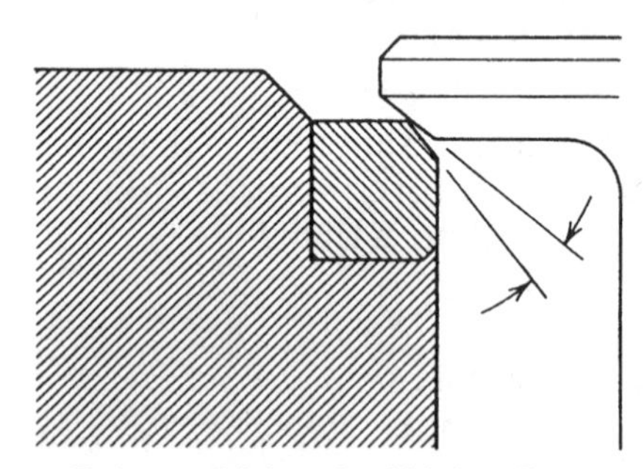

Exhaust Valve And Valve Seat

9. In the figure above, the angle indicated is called the:
   A. face angle.
   B. seat angle.
   C. interference angle.
   D. margin angle.
10. The LEAST likely cause for a diesel engine misfire is:
    A. poor fuel quality.
    B. bad injector.
    C. valve clearance out of specification.
    D. cracked cylinder sleeve.

11. A diesel engine will not turn crank when the key is moved to the start position. The LEAST likely cause of this problem is:
    A. bad battery connection.
    B. bad start switch.
    C. dead battery.
    D. fuel contamination.

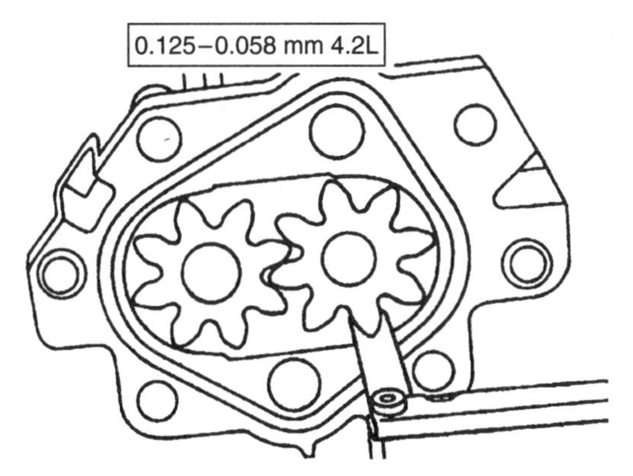

12. What type of measurement is being performed on the oil pump shown in the figure above?
    A. Radial
    B. End
    C. Axial
    D. Back-lash

Answers to the Test Questions for the Pretest

1. A, 2. B, 3. C, 4. B, 5. B, 6. D, 7. C, 8. C, 9. C, 10. D, 11. D, 12. A

Explanations to the Answers for the Pretest

Question #1
Answer A is correct because this process is not necessary when checking for cracks.
Answer B is wrong. A large variation in torque values is a good indicator of a potentially warped head, hence cracks.
Answer C is wrong. Finding small or otherwise invisible cracks can be greatly improved through the use of magnetic and dye type crack detectors.
Answer D is wrong. Sanding residue from head surface areas improves the clarity of the surface without introducing undesirable scrapes and scratches. Smaller cracks become visible as they are filled with particles from sanding.

Question #2
Answer A is wrong because this would cause a higher than normal or over temperature condition at high speed and load operation, generally leading to a constant overflow of fluid and formations of air pockets and hot spots. Indications of this condition would be most evident at the instrument panel until an air pocket formed around the temperature sensor(s), then the indication would drop to near zero.
Answer B is correct because a leaky fuel injector will not cause excessive surges in coolant level at normal temperatures.
Answer C is wrong because this condition would result in higher than normal coolant temperature after normal warmup. Progressive restrictions may lead to excessive cylinder head temp and then warping until it blows a head gasket.
Answer D is wrong because this condition will result in prolonged loss of fluid even when the engine is not hot.

Question #3
Answer A is a good choice because worn governor components will cause this condition. Yet, A is wrong because both technicians are right.
Answer B is a good choice because a clogged fuel filter will cause a low idle condition. Yet, B is wrong because both technicians are right.
Answer C is correct because a low idle could be caused by either worn governor parts or a clogged fuel filter.
Answer D is wrong because both technicians are right.

Question #4
Answer A is wrong because excessive smoke could be caused by a problem unrelated to the internal engine parts.
Answer B is correct because by discussing the problem with the owner or operator a correct diagnosis may occur.
Answer C is wrong because only one of the technicians is right.
Answer D is wrong because one of the technicians is right.

Question #5
Answer A is wrong because a dry sleeve does not contact engine coolant and therefore does not need an O-ring to seal it.
Answer B is correct because wet sleeve types use an O-ring to prevent coolant from leaking into the crankcase.
Answer C is wrong because a parent bore does not use sleeves of any type.
Answer D is wrong because "throw away" is a jargon term for a parent bore engine.

Question #6
Answer A is wrong because pitting if it is not too severe can be removed when refinishing the valves.
Answer B is wrong because if the valves are not damaged, they may not need to be replaced during an engine overhaul.
Answer C is wrong because it isn't necessary to replace the valves if you are resurfacing the seats.
Answer D is correct because if the valves show signs of cracking or warping, you replace them.

Question #7
Answer A is wrong because a piston sleeve may not need to be replaced due to normal wear.
Answer B is wrong because taper is considered to be normal wear.
Answer C is correct because any significant damage to the sleeve requires replacement and is the most likely reason.
Answer D is wrong because out-of-roundness is considered to be normal wear.

Question #8
Answer A is wrong because overloading must be very severe to cause bearing failure.
Answer B is wrong because corrosion is least likely in modern engines and rarely causes bearing failure.
Answer C is correct because dirt is the most common cause of bearing failure and therefore the most likely cause.
Answer D is wrong because misalignment is not a common cause of bearing failure.

Question #9
Answer A is wrong because the face angle is exclusive to the valve.
Answer B is wrong because the seat angle is exclusive to the seat.
Answer C is correct because the figure shows the interference angle.
Answer D is wrong because there is no such thing as a margin angle.

Question #10
Answer A is wrong because fuel quality is a common cause of engine misfires.
Answer B is wrong because a bad injector commonly causes engine misfires.
Answer C is wrong because valve clearance could cause an engine to misfire.
Answer D is correct because it is unusual for a sleeve to crack in a diesel engine.

Question #11
Answer A is wrong because bad connections are a common cause for no crank problems.
Answer B is wrong because a bad starter switch would cause a no crank condition to occur.
Answer C is wrong because a dead battery is a common cause for no crank problems.
Answer D is correct because even if fuel is contaminated, electrical current would flow allowing the engine to crank.

Question #12
Answer A is correct because the radial gear clearance is being checked in the figure.
Answer B is wrong because end clearance is not being shown.
Answer C is wrong because axial clearance is the same as end clearance and is not shown.
Answer D is wrong because back-lash is typically not measured in an oil pump.

Types of Questions

ASE certification tests are often thought of as being tricky. They may seem to be tricky if you do not completely understand what is being asked. The following examples will help you recognize certain types of ASE questions and avoid common errors.

Each test is made up of forty to eighty multiple-choice questions. Multiple-choice questions are an efficient way to test knowledge. To answer them correctly, you must think about each choice as a possibility, and then choose the one that best answers the question. To do this, read each word of the question carefully. Do not assume you know what the question is about until you have finished reading it.

Multiple-Choice Questions

One type of multiple-choice question has three wrong answers and one correct answer. The wrong answers, however, may be almost correct, so be careful not to jump at the first answer that seems to be correct. If all the answers seem to be correct, choose the answer that is the most correct. If you readily know the answer, this kind of question does not present a problem. If you are unsure of the answer, analyze the question and the answers. For example:

Question 1:

When the throttle is moved from wide open to closed the throttle position sensor should indicate:

A. 0 VDC to 5 VDC.
B. 5 VDC to 0 VDC.
C. 5 VDC to continuous.
D. 0 VDC to continuous.

Analysis:

Answer A is correct because zero to five volts is the standard voltage reference value on non-OBD II systems from WOT to the closed position.
Answer B is wrong because the voltage at full open is 0 and at full closed is 5.
Answer C is wrong because the voltage at full open is 0 and at full closed is 5.
Answer D is wrong because the voltage at full open is 0 and at full closed is 5.

EXCEPT Questions

Another type of question used on ASE tests has answers that are all correct except one. The correct answer for this type of question is the answer that is wrong. The word **EXCEPT** will always be in capital letters. You must identify which of the choices is the wrong answer. If you read quickly through the question, you may overlook what the question is asking and answer the question with the first correct statement. This will make your answer wrong. An example of this type of question and the analysis is as follows:

Question 2:

All of the following are used to determine battery capacity **EXCEPT:**

A. multimeter set to milliamps.
B. jumper wires attached to battery terminals.
C. accessory switch on.
D. battery.

Analysis:
Answer A is wrong. This is a required task when checking battery capacity.
Answer B is wrong. This is a required task when checking battery capacity.
Answer C is wrong. This is a required task when checking battery capacity.
Answer D is correct because the battery does not determine capacity, it stores it.

Technician A, Technician B Questions

The type of question that is most popularly associated with an ASE test is the "Technician A says . . . Technician B says . . . Who is right?" type. In this type of question, you must identify the correct statement or statements. To answer this type of question correctly, you must carefully read each technician's statement and judge it on its own merit to determine if the statement is true.

Typically, this type of question begins with a statement about some analysis or repair procedure. This is followed by two statements about the cause of the problem, proper inspection, identification, or repair choices. You are asked whether the first statement, the second statement, both statements, or neither statement is correct. Analyzing this type of question is a little easier than the other types because there are only two ideas to consider although there are still four choices for an answer.

Technician A . . . Technician B questions are really double-true-false questions. The best way to analyze this kind of question is to consider each technician's statement separately. Ask yourself, is A true or false? Is B true or false? Then select your answer from the four choices. An important point to remember is that an ASE Technician A . . . Technician B question will never have Technician A and B directly disagreeing with each other. That is why you must evaluate each statement independently. An example of this type of question and the analysis of it follows.

Question 3:

When an installed engine oil pressure gauge is compared to a test gauge at low engine rpm the installed gauge appears low, but at normal engine rpm, the installed gauge readings are correct. Technician A says that the oil passage in the block is clogged. Technician B says the line has a leak. Who is right?

A. A only
B. B only
C. Both A and B
D. Neither A nor B

Analysis:
Answer A is wrong. A clogged passage like a chinked line would result in slow response and faulty readings if the blockage is minimal, increasing until no reading is indicating.
Answer B is correct because the line has a leak when an installed engine oil pressure gauge is compared to a test gauge and at low engine the installed gauge appears low but at normal engine rpm the installed gauge readings are correct.
Answer C is wrong.
Answer D is wrong.

Questions with a Figure

About 10 percent of ASE questions will have a figure, as shown in the following example:

Question 4:

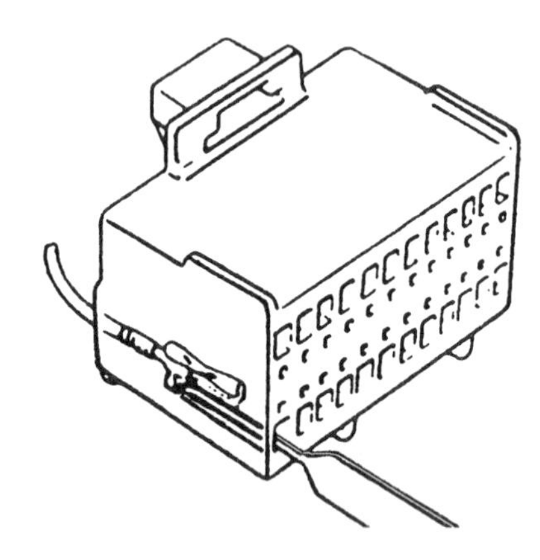

The figure above indicates the procedure for replacing which type of connector?

A. Micro
B. Weather pack
C. Pull-to-seal
D. Molded plug

Analysis:

Answer A is correct because the micro connector is shown.
Answer B is wrong.
Answer C is wrong.
Answer D is wrong.

Most-Likely Questions

Most-likely questions are somewhat difficult because only one choice is correct while the other three choices are nearly correct. An example of a most-likely-cause question is as follows:

Question 5:

Which tool is most likely to be used to determine the starter circuit voltage drop test?

A. Starter shunt resistor
B. Voltmeter
C. Ohmmeter
D. Milliammeter

Analysis:
Question #42
Answer A is wrong because this test is checking for excess resistance.
Answer B is correct because voltmeters measure voltage.
Answer C is wrong because this would damage the ohmmeter.
Answer D is wrong because this would damage the ammeter.

LEAST-Likely Questions

Notice that in most-likely questions there is no capitalization. This is not so with least-likely type questions. For this type of question, look for the choice that would be the LEAST likely cause of the described situation. Read the entire question carefully before choosing your answer. An example is as follows:

Question 6:

Which of the following is LEAST likely to happen?

A. A faulty injector may cause an engine to produce a white smoke.
B. Faulty or improperly installed oil control rings may result in the engine producing a white smoke.
C. A blown head gasket may result in the engine producing either a blue or white smoke, depending on where the gasket is blown.
D. Operating a diesel engine for extended periods in a low-load or no-load condition may result in the engine producing a black smoke or soot.

Analysis:
Answer A is wrong because excess unburned fuel that has not reached combustion temperature will produce white smoke.
Answer B is correct because faulty or improperly installed rings may result in the engine producing a white smoke.
Answer C is wrong because this condition will cause burning of oil or water that causes blue smoke and white smoke.
Answer D is wrong because this condition will cause low operating temperature and failing of the injectors that produce black smoke.

Summary

There are no four-part multiple-choice ASE questions having "none of the above" or "all of the above" choices. ASE does not use other types of questions, such as fill-in-the-blank, completion, true-false, word-matching, or essay. ASE does not require you to draw diagrams or sketches. If a formula or chart is required to answer a question, it is provided for you. There are no ASE questions that require you to use a pocket calculator.

Testing Time Length

An ASE test session is four hours and fifteen minutes. You may attempt from one to a maximum of four tests in one session. It is recommended, however, that no more than a total of 225 questions be attempted at any test session. This will allow for just over one minute for each question.

Visitors are not permitted at any time. If you wish to leave the test room, for any reason, you must first ask permission. If you finish your test early and wish to leave, you are permitted to do so only during specified dismissal periods.

Monitor Your Progress

You should monitor your progress and set an arbitrary limit to how much time you will need for each question. This should be based on the number of questions you are attempting. It is suggested that you wear a watch because some facilities may not have a clock visible to all areas of the room.

Registration

Test centers are assigned on a first-come, first-served basis. To register for an ASE certification test, you should enroll at least six weeks before the scheduled test date. This should provide sufficient time to assure you a spot in the test center. It should also give you enough time for study in preparation for the test. Test sessions are offered by ASE twice each year, in May and November, at over six hundred sites across the United States. Some tests that relate to emission testing also are given in August in several states.

To register, contact Automotive Service Excellence/American College Testing at:

ASE/ACT
P.O. Box 4007
Iowa City, IA 52243

4 An Overview of the System

Diesel Engines (Test T2)

The following section includes the task areas and task lists for this test and a written overview of the topics covered in the test.

The task list describes the actual work you should be able to do as a technician that you will be tested on by the ASE. This is your key to the test and you should review this section carefully. We have based our sample test and additional questions upon these tasks, and the overview section will also support your understanding of the task list. ASE advises that the questions on the test may not equal the number of tasks listed; the task lists tell you what ASE expects you to know how to do and be ready to be tested on.

At the end of each question in the Sample Test and Additional Test Questions sections, a letter and number will be used as a reference back to this section for additional study. Note the following example: **A15.**

Task List

A. General Engine Diagnosis (16 Questions)

Task A15 **Check lubrication system for contamination, oil level, temperature, pressure, filtration, and oil consumption; determine needed repairs.**

Example:

1. Technician A says to test the oil pressure gauge, it should simulate the instrument panel gauge in readings. Technician B says test readings should be taken during start up, varying operating ranges, and on shut down. Who is right?
   A. A only
   B. B only
   C. Both A and B
   D. Neither A nor B (A15)

Question #1
Answer A is a good choice because to test the oil pressure gauge, it should simulate the instrument panel gauge in readings. Yet A is wrong because both technicians are right.
Answer B is a good choice because test readings should be taken during start up, varying operating ranges, and on shut down. Yet B is wrong because both technicians are right.
Answer C is correct because both technicians are right.
Answer D is wrong because both technicians are right.

Task List and Overview

A. General Engine Diagnosis (16 Questions)

Task A1 Verify the complaint, and road/dyno test vehicle; review past maintenance documents (if available); determine further diagnosis.

The service technician must be aware of normal diesel engine operating noises in order to determine whether a system requires service. Normal noises include the sounds of diesel combustion, fuel injectors, etc.

The diesel engine technician needs to apply a basic diagnostic procedure such as a scientific process of elimination. This technician should perform the following diagnostic tasks:

- Listen carefully to the customer's complaint (concern), and question the customer to obtain more information concerning this complaint.
- Identify the complaint and road test or dyno the truck if necessary.
- Think of possible causes for this complaint.
- Work from what is working correctly on the truck.
- Perform diagnostic tests to locate the exact cause of the complaint. Always start with the easiest, quickest test, and then work toward the more difficult test.
- Be sure the customer's complaint is eliminated; road test or dyno the truck to verify the repair that eliminated the cause.

Task A2 Inspect engine assembly and compartment for fuel, oil, coolant, and other leaks; determine needed repairs.

To reduce airborne emissions, engine manufactures design engines to reduce significantly venting gases and fumes. Most of the remaining gases and fumes are recirculated through the combustion chamber. These efforts have also produced a cleaner engine and engine compartment. Careful inspection of the engine and engine compartment can reveal potential problems.

For example: Oil residue on the bottom side of the hood may indicate excess crankcase pressure caused by blowby or an overfilled crankcase. Oil on the exhaust could be a leaking head gasket or a bad valve cover gasket.

Stains on the engines that appear discolored or lighter than the surrounding area could be from a small coolant leak from the hose or the head. A light, slippery, almost invisible coating on most of the engine compartment surfaces could be caused by a small leak in the radiator. This could be from excess venting of radiator pressure or a very small leak in a high-pressure fuel or hydraulic line.

Deposits of liquid in cracks and crevices or low spots can be indicators of leaks from the surrounding components, lines, and/or hoses. Leak detection is a common-sense skill. Remember:

Airflow is usually from front to rear of the engine compartment during truck operation. Gravity causes liquids to accumulate at the lowest point. If fluids are lost, they have to go somewhere, usually downward or out of the exhaust.

Task A3 Inspect engine compartment wiring harness, connectors, seals, and locks; determine needed repairs.

Modern engine electrical/electronic wiring is more complex and critical. If connectors or wires are bad, the circuit sends faulty inputs from the monitoring sensors that operate on milliamps of current. This, in turn, causes computerized circuits using these inputs to make faulty adjustments to the engine control. Some of the common problems associated with circuits are corrosion and other foreign substances in connectors, loose or improperly mated connectors, and improper grounding. Some of these problems will not affect engine operation until several other problems exist, but then they confuse the solution process. If a visual inspection of a connector reveals corrosion, the connector should be cleaned or replaced. One needs to repair wiring that is visibly lumpy or has nicked insulation to prevent faults from corrosion. Loose or scale coated grounds should be removed, cleaned, treated for corrosion, reinstalled, and coated with a corrosion inhibitor. All required connections to a ground must be free of all rust and/or paint to ensure a good electrical connection.

Task A4 Listen to engine noises; determine needed repairs.

A technician who develops a discriminating ear for engine noises can simplify troubleshooting or find problems before they are indicated in diagnostics. Listening though a metal bar or heater hose applied to each cylinder in turn can expose a misfiring cylinder like a stethoscope. Sometimes the absence of change in the sound may provide confirmation of a faulty condition. Again, if one of the cylinders appears to be misfiring, manually foul each injector. Every time you foul a good cylinder, the sound of the engine should change. On the other hand, a fouled cylinder will have no noticeable change in sound. Often the frequency of an unusual noise is relevant to the defective component's operating speed. A loud knocking that increases at the same rate as the engine would probably be associated with the crank. Whereas a sound that is 1/2 the engine speed might be created by a cam or something directly attached to it. Other components may produce high frequency sounds like a squeal, and may not be as discernible using this method. In these cases, turning your head from side to side to determine the direction of loudest sound may point to the general location. Then, look for components in that area that operate at a high speed. Squeals may also be produced when two surfaces are rubbing each other, like belts slipping. Sometimes sounds appear random in nature. Usually the operator, an automated operation, or a load change, like a power steering pump squealing when the steering wheel is turned causes these random changes.

Task A5 Check engine exhaust emissions, odor, smoke color, opacity (density), and quantity; determine needed repairs.

The color and consistency of the exhaust can reveal problems associated with the combustion of the engine. When a diesel engine is running properly with a heavy load, its exhaust is almost clear. The only indication of the exhaust is a slight distortion in the objects viewed through the exhaust. As a rule, when chemicals are burnt or suspended in the exhaust they produce a discoloration or smoke. Because of variations in engine operation, such as changes in fuel demand, loading, and outside temperature, some degree of smoke is expected. Excessive or prolonged appearance of smoke can be an indicator of other problems. Oil, for instance, adds a blue haze to the exhaust, causing a blue smoke. Unburned fuel or an improper grade of fuel produces a gray or black smoke; if this becomes excessive, soot will form. This could also occur if the intake system has provided insufficient air for fuel delivered. Water or coolant will produce a white smoke-like steam. White smoke may also indicate that injectors are misfiring, this happens when combustion does not occur and fuel vapors are present in the exhaust. White smoke occurs either when the engine starts the first time or when the ambient temperature is too low for proper combustion.

Because of incomplete mixing of the injected fuel with the air in the combustion chamber, some fuel droplets do not burn until late in the cycle. This means that the combustion chamber temperature is lower and there is also less oxygen to sustain the remaining burn; therefore, incomplete combustion occurs. This situation results in smoke coming out the exhaust. White smoke results from incomplete combustion in overlean combustion chamber areas, by fuel spray impingement on metal surfaces, and with low temperatures in the cylinder, such as when starting an engine (more so on cold days). Fuel with too high a cetane rating can also cause white smoke when used in high-ambient-temperature conditions. White smoke occurs more often in an indirect injected

(IDI) engine from retarded timing. Very late timing on a direct injected (DI) engine causes white smoke. Gray or black smoke is the result of incomplete combustion in rich combustion chamber areas, caused by such conditions as engine overload, insufficient fuel injector spray penetration, late ignition due to retarded injection timing, or poor fuel evaporation and mixing due to advanced timing. Air starvation is the major cause of black smoke.

Note: MIXING: Owing to the extremely short time available for mixing, as the fuel-air rate increases beyond a certain value, an appreciable fraction of the fuel fails to find the necessary oxygen (O_2) for combustion and passes through the cylinder unburned or partially burned.

Task A6 Perform fuel system tests; check fuel consumption; determine needed repairs.

If the engine runs but is low on horsepower, examine the following items to detect the problem: Check the fuel filter and change it if necessary.

Check the snap pressure by attaching a pressure gauge to an extra outlet fitting on the shutdown solenoid and then accelerating the engine quickly from idle to full throttle. Read the maximum pressure on the gauge during acceleration. It should be very near the pressure listed on the pump calibration sheet under engine fuel pressure. Snap pressure checks will not be accurate on AFC pumps.

If the snap pressure is low, it may be necessary to change the throttle restriction or the idle spring plunger button. If there is an incorrect high idle setting, change the governor spring shims. If the throttle travel is not correct, check to make sure the throttle is in the wide-open position with the accelerator pedal all the way down.

Maximum delivery check: Before making this check back out the torque screw and high idle screw approximately five turns. Run the pump at speed specified for full (rated) fuel and divert fuel flow to graduates. Collect fuel for 1,000 strokes. Run at least three tests before recording the delivery. This wets the glass of the graduates and removes air from the lines. If the delivery is incorrect, it must be adjusted by removing the governor cover and main governors spring and guide stud.

Rotate the pump until the leaf spring adjusting screw can be seen through the cam ring. With an Allen wrench, adjust the leaf spring. Replace the parts removed and recheck the fuel delivery.

Task A7 Perform air intake system restriction and leakage tests; determine needed repairs.

When an engine suffers from a loss of power or rough operation, you should go back to the basics. To obtain the optimum performance from an engine the proper quantity and quality of fuel must be supplied to the cylinder at the right time with the right amount of air and heat.

Air starvation will generally result in incomplete combustion, increasing black smoke or soot, a decrease in power, and heat build up in the cylinder due to insufficient scavenging. Some of the reasons for air starvation are restricted intake hoses and pipes, dirty air filter, excessive exhaust backpressure, or high altitude operation. When the diesel engine is equipped with a blower or turbocharger, check for leaks in the output tubes and the aftercooler. Verify that the aftercooler is functioning properly. Ultimately, the denser the air, the better the engine runs. Fuel starvation will generally result in total combustion of fuel supplied with a loss of power, no noticeable change in exhaust color, and low cylinder temperatures. Some of the reasons for fuel starvation are an empty tank, defective fuel pump, leaks in fuel lines from the tank, restricted filters or strainers, and crimped fuel lines between the pump and injectors. A temperature spread of less than 10°F (6°C) between the fuel's pour point and temperature can restrict fuel flow enough to cause starvation. Faulty injectors or improper spray patterns can also reduce power and cause erratic operation; this is usually accompanied with black or gray smoke.

Gray or black smoke is the result of incomplete combustion in rich combustion chamber areas, caused by such conditions as engine overload, insufficient fuel injector spray penetration, late ignition due to retarded injection timing, or poor fuel evaporation and mixing due to advanced timing. Air starvation is the major cause of black smoke.

Task A8 Perform manifold pressure and/or air box pressure tests; determine needed repairs.

The crankcase pressure indicates the amount of air passing between the oil control rings and the cylinder lines into the crankcase, most of which is clean air from the air box. A slight pressure in the crankcase is desirable to prevent the entrance of dust. A loss of engine lubricating oil through the breather tube, crankcase ventilator, or dipstick hole in the cylinder block is indicative of excessive crankcase pressure. The causes of high crankcase pressure may be traced to excessive blowby due to worn piston rings, a hole or crack in a piston crown, loose piston pin retainers, worn blower oil seals, defective blower, cylinder head gaskets, or excessive exhaust backpressure. In addition, the breather tube or crankcase ventilator should be checked for obstructions. Check the crankcase pressure with a manometer connected to the oil level dipstick opening in the cylinder block. Check the readings obtained at various engine speeds with the manufacturer manual.

Task A9 Perform exhaust backpressure and temperature tests; determine needed repairs.

A slight pressure in the exhaust system is normal. However, excessive exhaust backpressure seriously affects engine operation. It may cause an increase in the air box pressure with a resultant loss of efficiency of the blower. This means less air for scavenging, which results in poor combustion and higher temperatures. Causes of high exhaust backpressure are usually a result of an inadequate or improper type of muffler, an exhaust pipe which is too long or too small in diameter, an excessive number of sharp bends in the exhaust system, or obstructions such as excessive carbon formation or foreign matter in the exhaust system. Check the exhaust backpressure, measured in inches of mercury, with a manometer. Connect the manometer to the exhaust manifold (except on turbocharged engines) by removing a pipe plug which is provided for that purpose. If no opening is provided, drill an 11/32 in. hole in the exhaust manifold companion flange and tap the hole to accommodate a pipe plug. On turbocharged engines, check the exhaust backpressure in the exhaust piping 6 to 12 in. from the turbine outlet. The tapped hole must be in a comparatively straight pipe area for an accurate measurement.

Task A10 Perform crankcase pressure test; determine needed repairs.

The crankcase pressure indicates the amount of air passing between the oil control rings and the cylinder lines into the crankcase, most of which is clean air from the air box. A slight pressure in the crankcase is desirable to prevent the entrance of dust. A loss of engine lubricating oil through the breather tube, crankcase ventilator, or dipstick hole in the cylinder block is indicative of excessive crankcase pressure. The causes of high crankcase pressure may be traced to excessive blowby due to worn piston rings, a hole or crack in a piston crown, loose piston pin retainers, worn blower oil seals, defective blower, cylinder head gaskets, or excessive exhaust backpressure. In addition, the breather tube or crankcase ventilator should be checked for obstructions. Check the crankcase pressure with a manometer connected to the oil level dipstick opening in the cylinder block. Check the readings obtained at various engine speeds with the manufacturer manual.

Task A11 Diagnose no cranking, cranks but fails to start, hard starting, and starts but does not continue to run problems; determine needed repairs.

Often when an engine cranks but fails to start it is an indication of air in the fuel system. To determine if air is causing the problems pressurize the system using a hand primer and bleed the lines to each injector while cranking the engine. If air is found, examine the area around the fuel lines between the primer and injectors for an accumulation of fuel. If the primer pump fails to build up pressure within the required time, check fuel filter lines for leaks and the tank for adequate fuel level.

A diesel engine may not start or be hard to start because the temperature of the air in the cylinder is too low for full combustion of the atomized fuel. If the starter does not rotate the engine fast enough to allow the air in the cylinders to reach combustion temperature, combustion will not occur.

Electrically, the starter may turn slowly if it is defective, has a low battery potential, has corroded cables, or an improper electrical connector. Inspect the starter, battery, and connections to ensure proper operation of the starter motor.

Mechanically, the engine may offer too large a load on the starter. This happens when the oil is too thick or the surfaces too cold, the belt or gear driven devices are loading down the engine, or there is too much cylinder backpressure. Check oil specifications for that climate zone, check immersion heaters if installed, check for proper belt tension and gear backlash, and use unloader valves if installed.

- Low combustion temperature may also result if the outside temperature is too low and you do not use preheating.
- To assist in starting the engine at low temperatures use ether, which will reduce the flash point of the fuel/air mixture.
- To maintain a warm engine for starting at low temperatures the engine may use an immersion heater in the cooling system or submerged in the crankcase oil.

A reduction in compression ratio created by a defective turbo or blower, a restricted air intake or excessive exhaust backpressure will also delay or prevent combustion pressure build up.

Task A12 Diagnose surging, rough operation, misfiring, low power, slow deceleration, slow acceleration, and shutdown problems; determine needed repairs.

When an engine suffers from a loss of power or rough operation, you should go back to the basics. To obtain the optimum performance from an engine the proper quantity and quality of fuel must be supplied to the cylinder at the right time with the right amount of air and heat.

Air starvation will generally result in incomplete combustion, increasing black smoke or soot, a decrease in power, and heat build up in the cylinder due to insufficient scavenging. Some of the reasons for air starvation are restricted intake hoses and pipes, dirty air filter, excessive exhaust backpressure, or high altitude operation. If the diesel comes equipped with a blower or turbocharger one needs to check for leaks in the output tubes and the aftercooler. Verify that the aftercooler is functioning properly. Ultimately, the denser the air, the better the engine runs. Fuel starvation will generally result in total combustion of fuel supplied with a loss of power, no noticeable change in exhaust color, and low cylinder temperatures. Some of the reasons for fuel starvation are an empty tank, defective fuel pump, leaks in fuel lines from the tank, restricted filters or strainers, and crimped fuel lines between the pump and injectors. A temperature spread of less than 10°F (6°C) between the fuel's pour point and temperature can restrict fuel flow enough to cause starvation. Faulty injectors or improper spray patterns can also reduce power and cause erratic operation; this is usually accompanied with black or gray smoke.

Task A13 Isolate and diagnose engine related vibration problems; determine needed repairs.

Often the frequency of occurrence of an unusual vibration is a direct or multiple of the operational speed of the device creating the vibrations. A strong vibration that increases occurrence rate with engine speed would probably be the crank or something directly attached to it. Whereas a vibration that has an occurrence rate 1/2 the engine speed might be created by a cam or something directly attached to it. Other components, like an out of balance turbocharger or alternator, may produce high frequency vibrations that seem to have a constant occurrence rate. In these cases, use a metal bar like a stethoscope to find the strongest concentration of that vibration. Then, look for components in that area that operate at a high speed. High frequency vibrations can also be produced by bearings that are failing or two surfaces that are rubbing each other. These are usually destructive in nature, have a constant occurrence rate, progressively grow in intensity, and last for a short time before they become more apparent. Some automated operations, like a cycling air condition, may have a vibration occurrence rate that seems random. Two or more vibrations of different frequency occupying at the same time may produce a rhythmic vibration.

Task A14 Check cooling system for protection level, contamination, coolant level, temperature, pressure, conditioner concentration, filtration, and fan operation; determine needed repairs.

The radiator cap will maintain the pressure within the cooling system at or below approximately 15 psi (103 kPa) During heavy loads and prolonged periods of idling, this pressure may exceed the rating of the radiator cap and fluid will be vented to an expansion tank or bottle. Once enough fluid has been vented to reduce the pressure within limits, the spring tension of the radiator cap will close the vent line. At normal temperatures, with the pressure cap removed, small fluctuations in coolant level can be expected. Inspection of this fluid level may indicate deeper problems. A constant overflowing of the fluid from the tube may indicate air pockets, uneven heating, or hot spots in the engine. Repetitious and rhythmic excessive surges in this level usually indicate a leak to the combustion cylinder, such as a blown head gasket.

At different and spontaneous times, the cooling system will overheat. At the time, the system should be checked as a unit in this manner: check tightness on hose clamps, check for soft or sometimes mushy feeling hoses and check for cracks. You should be able to see all the components and look for leaks, such as at the radiator, water pump, cooling filter, head gaskets, and water manifold. You should be able to notice a red or greenish color, or a residue that builds up at the leak. Belts and the fan should be checked for aging. Pressurize the system and watch hoses, radiator tank joints, and head gaskets for possible points of leakage. Test lubrication oil for water and obstructions at the radiator and fins. Remove and test the thermostat for proper opening and closing temperature, and replace as necessary. If the thermostat is found defective, usually it is an indication of something else causing an overheating condition. Usually, thermostats last a long time, unless the engine exceeds the operating range of the thermostat.

Task A15 Check lubrication system for contamination, oil level, temperature, pressure, filtration, and oil consumption; determine needed repairs.

Examination of lubrication oil can reveal problems associated with engine performance. An indication of oil dilution from coolant is water droplets forming on the dipstick while the engine is off, or if the oil appears gray in color after the engine has run. This can occur for any of the following; a cracked block or cylinder head, blown head gasket, leaky oil cooler, or a leaky sleeve O-ring. Fuel in the oil will cause the oil to thin out when cold and may appear on the dipstick as clearer oil at top of level indicator if the engine has not been run for a while. This problem may be caused by leaky injectors or a leaky injector pump.

Excessive oil consumption can be a good indicator of improperly seated or defective rings, a damaged piston, or defective valve stem seals. These problems will usually be accompanied by a blue or black smoke associated with the exhaust. Low oil pressure may indicate a worn oil pump, restricted suction tube or screen, open, or improperly adjusted relief valve, clogged oil filter, or worn bearings. Regardless of the initial problem, if it is not corrected, problems that are more serious will appear.

Task A16 Check, record, and clear electronic diagnostic codes; monitor electronic data; determine needed repairs.

Today's diesel engines contain a variety of computerized engine control systems depending on the engine manufacturer. According to SAE Standard J1930, diagnostic codes are known as Diagnostic Trouble Codes or DTCs. The DTC extraction method varies with the engine and truck manufacturer. You need to understand how to check, record, and clear DTCs. The typical process is to use a scan tool, diagnostic reader, or laptop computer to check for, record, and clear DTCs.

Modern engines are controlled and monitored by electronic modules. Some of these modules are capable of storing error codes against engine parameter identifications (PID). To retrieve these codes, the technician connects a diagnostic test set to a diagnostic plug. Some vehicles will display a limited readout of PIDs by blinking an instrument panel indicator in a two-digit code called a flash code. Diagnostic test sets also have a menu driven system and module check that can be run. To test the engine, once the problems have been analyzed and corrected, the PID codes need to be cleared. Diagnostic test sets can be used to clear these codes. Another approach is to disconnect the battery.

B. Cylinder Head and Valve Train Diagnosis and Repair (8 Questions)

Task B1 Remove, inspect, disassemble, and clean cylinder head assembly(s).

Inspection of the cylinder head begins by checking the torque of cylinder head bolts before removing the head. This can indicate the potential for a warped or cracked head. Inspect the head gasket and head surface for signs of carbon or coolant stains between cylinders and passages before steam cleaning the head. Use a steam cleaner to remove oil, coolant, and carbon residue. Then, sand or peen the surface as necessary to remove all remaining carbon and gasket material. Use of sharp objects like knives or scrappers to remove residue may impair inspection for cracks and damage. When cracks are suspect, but not apparent, magnetic or dye crack detection may be required to highlight cracks. Visually re-inspect the head for cracks in the casting, scale buildup in the coolant passages, and pitting of machined surfaces. Excess accumulation of scale in water passages can usually be removed by placing the head in a non-corrosive dip for about two hours. Pressure testing the head can be used to locate a crack in coolant and oil passages.

Task B2 Inspect threaded holes, studs, and bolts for serviceability; service/replace as needed.

When removing or installing a component, carefully inspect all bolts, studs, holes, and nuts for damage to threads. Stripped, cross-threaded, nicked, rolled, and rusty threads can produce errors in the torque values. Use compressed air to remove foreign matter from holes. Obstructions lodged in holes, such as metal and liquids, can cause a bolt to bottom out and reach torque value without obtaining the desired results. Examine the shank of bolts and studs for signs of twisting or over-torque fracturing. Replace defective bolts and studs. Ensure that nuts are not distorted or cracked, replace self-locking and defective nuts. Clean and redress threads in holes. If threads in a hole are damaged beyond use, drill and tap the hole and install an approved helicoil. Use an approved stud extractor or Eze-out to remove broken studs and bolts.

Task B3 Measure cylinder head deck-to-deck thickness, and check mating surfaces for warpage; inspect for cracks/damage; check condition of passages; inspect core, gallery, and plugs; service as needed.

The cylinder head should be checked for warping. Using a straightedge and feeler gauge, measure the face (fire deck) for flatness. Check for transverse (across the width of a cylinder head) warpage at each end and between each cylinder. Check for longitudinal (across the length of the cylinder head) warpage at points above and below each cylinder, between center centerlines of valves, and the outer edge of the fire deck. Compare measurements to manufacturer's specifications to determine the need for resurfacing. Resurfacing may necessitate removing inserts, such as valve guides, and/or deburring of water ports. Check cylinder head thickness, called deck-to-deck thickness, to determine the maximum resurfacing allowable.

Task B4 Test cylinder head for leakage using approved methods; service as needed.

When manufacturers incorporate fuel delivery passages into a cylinder head, use their procedures to pressurize with air and test for head-fuel leaks. Inherently, air will indicate the presence of a leak that liquids will not. This method does not introduce future contamination. Often this procedure requires installing the discarded injectors or a jig, sealing the return port with a threaded plug, and using a regulated air supply. Water passages are checked using a combination of air pressure and water heated to operating temperature. To ensure satisfactory results when performing these tests follow the manufacturer's procedures and use only jigs and tools specified. Usually this requires installation of specialized jigs to seal coolant ports and provide connections for coolant and air supply.

Task B5 Inspect and test valve springs for squareness, pressure, and free height comparison; replace as needed.

Check valve spring straightness with a T-square. This will help prevent stems from binding and twisting in guides as valves are operated. This uneven side pressure causes the valve stems and guides to wear faster. Also check the unloaded, or free length, and loaded length, or tension, of the spring using a spring tension gauge. Replace valve springs that do not meet specifications, to prevent damage to the piston or crankshaft. Valve springs using the same valve bridge must be replaced together to prevent unbalanced valve operation. This can cause damage to the bridge from uneven stress, valve guides and stems to wear faster, and insufficient scavenging of the cylinder.

Task B6 Inspect valve spring retainers and/or rotators, locks, and seals; replace as needed.

Valve spring retainers keep the valve spring on the valve, and valve rotators are used to rotate the valve each time it is used. Some engines also have seals. Careful inspection of these components can prevent serious damage and extend the life of the engine. Retainers, also called keepers that become worn or damaged, may allow the spring to release the valve. If this happens, the valve drops into the cylinder and jams the piston, called dropping a valve. Free release rotators use engine vibration and exhaust gases to rotate, because the spring tension is removed when it is depressed. Mechanical rotators apply a torque to rotate the valve each time the valve is opened. Both will eliminate carbon deposits from forming on the valve face and seats. Seals between the valve guide and valve stem prevents oil from entering the cylinder on the intake stroke. Defective seals can cause oil consumption, leading to excess smoking, carbon buildup, fueling of injectors, formation of hot spots, and precombustion or dieseling.

Task B7 Measure valve guides for wear, check valve guide-to-stem clearance, and measure valve guide height; recondition/replace as needed.

After the cylinder head has been checked or resurfaced and is considered useable, the valve guides should be checked for wear as follows:

- Check the guide inside diameter with a snap, ball, or dial gauge in three different locations throughout the length of the guide.

If guides are worn excessively, they should be replaced, if they are the replaceable types. In most cylinder heads, the valve guides can be removed by using the driver. Select a driver that fits the valve guide. After selection of a driver the guide should be removed as follows:

- Support the cylinder head on a block or stand to allow one to drive out the guide.
- If guide is to be pressed out, place the cylinder head in the press and align the driver with the press ram.
- Note the position of the guide in relation to the cylinder head for reference when installing a new guide.
- Drive or press the guide out of cylinder head making sure that the driver is driven or pressed straight. If it is not pressed straight, damage to the valve guide or cylinder head may result.
- After the guide has been removed, check the guide bore for scoring. A badly scored guide bore may have to be reamed out to accommodate the next larger size guide.
- After the guide bore is checked, select the correct guide (intake or exhaust) and insert it in the guide bore.
- Insert the guide driver and drive or press the guide into the cylinder head until the correct guide position is reached. If manufacturer's specifications are not available, position the guide in the same position of the old guide.
- Many engine manufacturers recommend that a new guide be hand-reamed after installation to ensure that the guide inside diameter did not change during the installation process.

Task B8 Inspect, recondition, or replace valves.

A common practice during major maintenance is to replace all valves if they have been in the head for several thousand operating hours. Valves must be changed if the head shows signs of carbon buildup from oil leaking around the valves. In this case, also consider changing valve guides. Valves with scratched, scored, pitted, or worn stems or stems with worn keeper grooves should be replaced. If the valve head is cupped, nicked, or marred it should be replaced. If a valve passes the visual inspection, it must be cleaned using a wire buffing wheel or glass bead blaster, if available. Use a valve refacing tool or valve grinder to obtain the correct face angle. Inspect the valve to ensure that the valve margin is within specifications. Insufficient valve margin may result in premature burning and subsequent failure. Finally, valve stem ends are ground flat and a small taper, usually about 1/32 to 1/16 in. (0.79 to 1.59 mm) is applied.

Task B9 Inspect, recondition, or install/replace valve seats.

The valve seats are used to absorb wearing of the head caused by valves opening and closing, air or heated exhaust gases flowing, and the effects of expansion and contraction. It is important to properly inspect, clean, and recondition or replace valve seats as necessary to maintain a good valve to head seal. Loose valve seats are located by lightly tapping each seat with the peen end of a ball peen hammer. A loose seat will produce a sound that differs from the ring sound created by tapping the head adjacent to the seat. The seat may even move a little. Leaks may occur if the seat has visible cracks, uneven burns or discoloration, or an opening which is oblong or excessive in diameter. Replace any seats that cannot be reconditioned. To recondition a valve seat, a mandrel pilot is installed in the valve guide to keep a specially dressed grinding wheel centered in the seat. The grinding wheel may be dressed with a one-degree steeper angle than the valve. This allows the valve to cut through future carbon buildup. The one-degree difference is not necessary if valves are using rotators. New seats must be reconditioned for the valve that will use them. Repeat the inspection process after reconditioning the valve seats. Also, check the concentricity (roundness) of the seat opening. If the seat is not round, pressure may escape during compression and combustion strokes. Verify that the valve head height is correct. If the head height is out of specification seats may need to be replaced, as this may alter the compression ratios or injector spray pattern. It is important not to remove any more material than is necessary to refinish the valve seat.

After guides have been replaced or reconditioned, valve seat checking and reconditioning should be done. Valve seats must be checked for looseness by tapping the seat lightly with the peen end of a ball peen hammer. A loose seat will produce a sound different from the sound produced while tapping on the cylinder head. In some cases, a loose seat can be seen to move while tapping on it. If the seat is solid, check it for cracks and excessive width. Normal valve seat width is 0.0625 to 0.125 in. (1.59 to 3.175 mm). If the seat passes all checks it should be reconditioned as outlined, using a specially designed valve seat grinder. Most valve seat grinders are similar to the one shown in Figure 4-1. Select the proper mandrel pilot by measuring the valve stem with a micrometer or caliper or by referring to manufacturer's specifications. After experience has been gained in this area, selection of the pilot is easily done by a visual check. Pilots are usually one of two types: expandable or tapered. An expandable pilot shown in Figure 4-2 is inserted into the guide and then expanded, tightening the expanding screw on the pilot, which expands the pilot. Expandable pilots are not considered as accurate as tapered pilots and should be used only when a tapered pilot is not available. Some mechanics prefer expandable pilots because they compensate for guide wear better than a tapered pilot; by expanding the guide the pilot tightens and adjusts the guide size or wear. A tapered pilot does not have an expanding screw and relies on the taper of the pilot to tighten it into the guide.

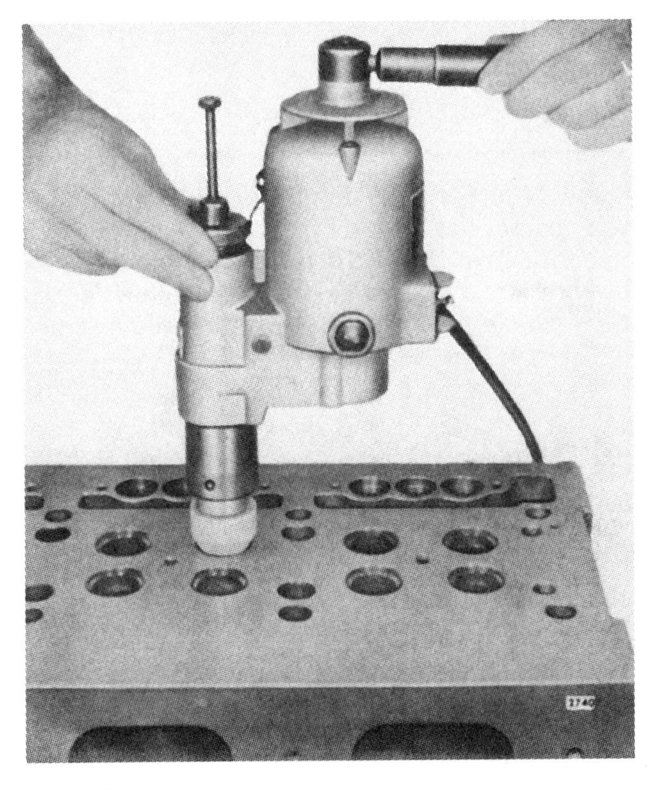

Figure 4-1

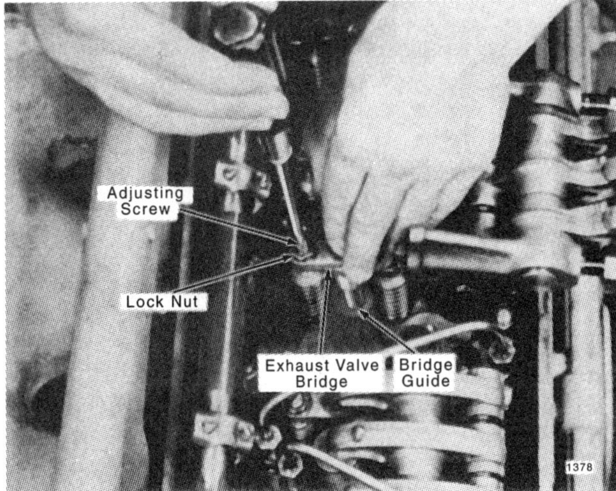

Figure 4-2

Task B10 Measure valve head height relative to deck, valve face-to-seat contact, and valve seat concentricity.

Before installing valves, check the valve to seat contact. One method of doing this is to coat the face of the valve with Prussian Blue and lightly snapping the valve into its seat. Remove the valve and check its face. If the valve face does not have an even seating mark, the seat is not concentric and must be reground. Another way to check contact is to place pencil marks around the face or the valve about 1/8 in. apart (3 mm) and snaps the valve against its seat. Remove the valve and check the face. If any of the marks are unbroken, the valve seat must be reground. When the valves have been seated, check the valve heads height. This checks the distance the valve protrudes above or below the machined surface of the head. Place a straightedge across the cylinder head and use a feeler gauge to measure distance between the valve head and straightedge. If the valve is designed to protrude above the head, place the straightedge on the valve. Use a feeler gauge between the valve head and the cylinder head machined surface.

Task B11 Inspect and replace injector sleeves and seals; measure injector tip or nozzle protrusion where specified by manufacturer.

If a leak is detected in an injector sleeve seal, the seal must be replaced to prevent further damage to the engine. After performing the manufacturer's procedures for removing and reinstalling the injector sleeve, perform a cylinder head pressure test to ensure the sleeve is sealed. Check the injector tip height to ensure it meets manufacturer specifications. Failure to do so may result in improper combustion due to injector spray pattern striking the cylinder walls or valves instead of the piston.

Task B12 Inspect, clean, and/or replace precombustion chambers where specified by manufacturer.

On IDI applications, precombustion chambers are either integral or installed as units in the head. You install a precombustion chamber into the cylinder head with a plastic hammer flush to plus .002 in. (.050 mm).

Task B13 Inspect and/or replace valve bridges (crossheads) and guides; adjust bridges (crossheads).

Valve crossheads are used on some types of diesel engines that use "four valve heads." Four valve heads used on four stroke cycle engines have two intake and two exhaust valves per cylinder. Two stroke cycle engines, such as Detroit Diesel, use four exhaust valves in each cylinder. The crosshead is a bracket or bridge-like device that allows a single rocker arm to open two valves at the same time. The crossheads (bridge) must be checked for wear as follows:

- Visually check crosshead for cracks and with magnetic crack detector if available.
- Check crosshead inside diameter for out-of-roundness and excessive diameter.
- Visually check for wear at the point of contact between the rocker lever and the crosshead.
- Check the adjusting screw threads for broken or worn threads.
- Check the crosshead guide pin for diameter with a micrometer.
- Check the crosshead guide pin to ensure that it is at right angles to the head-milled surface.

Task B14 Clean components; reassemble, check, and install cylinder head assembly as specified by the manufacturer.

The assembly of the cylinder head or heads onto the engine block will involve many different procedures that are peculiar to a given engine. The following procedures are general in nature and are offered to supplement the manufacturer's service manual.

Before attempting to install the cylinder head, make sure the cylinder head and block surface are free from all rust, dirt, old gasket materials, and grease or oil. Before placing the head gasket on the block make sure all bolt holes in the block are free of oil and dirt by blowing them out with compressed air.

Select the correct head gasket and place it on the cylinder block, checking it closely for an "up" or "top" mark that some head gaskets may have. In most cases head gaskets will be installed dry with no sealer, although in some situations an engine manufacturer may recommend applying sealer to the gasket before cylinder head installation. Take note of the recommendations in the service manual or ask your instructor.

After placing the head gasket on the block, place water and oil O-rings (if used) in the correct positions. Some head gaskets will be of a one-piece solid composition type, while others will be made of steel and composed of several sections or pieces. Detroit Diesel, for example, uses a round circular tin ring that fits on top of the cylinder sleeve to seal the compression. In addition to this sleeve, seals are the compression. In addition to this sleeve, seals are numerous O-rings that seal the coolant and lubrication oil, making up the head gasket. The installation of some head gaskets requires the use of a dowel (threaded rods) that are screwed in the head bolt holes to hold the head gasket in place during head installation. If dowels are recommended, they can be made from bolts by sawing off the heads and finding a taper on the end. Please insert correct word. Most modern engines will have dowels or locating pins in the block to aid in holding the cylinder head gasket in place during cylinder head assembly.

After the gasket and all O-rings are in place, check the cylinders to make sure no foreign objects have been left in them, such as O-rings and bolts.

Place the head or heads on the cylinder block carefully to avoid damage to the head gasket. On in-line engines with three separate heads that do not use dowel pins, it may be necessary to line up heads by placing a straightedge across the intake or exhaust manifold surfaces.

Clean and inspect all head bolts or cap screws for erosion or pitting. To clean bolts that are very rusty and dirty use a wire wheel. Coat bolt threads with light oil or diesel fuel and place the bolts or cap screws into the boltholes in the head and the block.

Using a speed handle wrench and the appropriate size socket, start at the center of the cylinder head and turn the bolts down snug, which is until the bolt touches the head and increased torque is required to turn it.

When all bolts have been tightened this far, continue to tighten each bolt one-quarter to one-half turn at a time with a torque wrench until the recommended torque is reached.

Clean the valve rocker covers and around the covers before removing them from the engine to avoid dust or dirt from entering the valve mechanism. Then, loosen the bolts (current engines) or knobs (former engines) and lift each cover straight up from the cylinder head. Use new gaskets when reinstalling the covers. Before a diecast rocker cover is installed on a cylinder head, it is important that the silicone gasket be properly installed in its groove in the rocker cover.

Clean and blow out the groove in the rocker cover with compressed air. Oil in the rocker covers groove or on the silicone gasket will make it difficult to install.

Press the stem side of the new T-shaped gasket down into the groove at the four corners of the cover first. Then, press the remainder of the gasket into place in the groove. Be sure the stem of the entire gasket bottoms in the groove.

Before installing the rocker cover, lubricate the cylinder head rail and the flat surface of the gasket with a thin film of engine oil. This will keep the gasket from sticking to the cylinder head rail.

Task B15 Inspect pushrods, rocker arms, rocker arm shafts, electronic wiring harness, and brackets for wear, bending, cracks, looseness, and blocked oil passages; repair/replace as needed.

After the cylinder heads have been tightened to the correct specification, the push rods or tubes and rocker arm assemblies may be installed. Straightening of push tubes is discouraged. Bent ones should be replaced. Push tubes must be checked for breaks or cracks where the ball socket on either end had been fitted into the tube. Place rod or tubes in the engine, making sure that they fit into the cam followers or tappets. The rocker arm assembly or rocker arm is one component part of the engine that is occasionally overlooked during a diesel engine overhaul. Some mechanics have the mistaken assumption that rocker arms wear very little or not at all. This is not true and rocker arm wear may account for increased engine oil consumption. Oil consumption occurs because the increased clearance allows an excessive amount of oil to splash or leak on the valve stem. This oil will run down the valve and end up in the combustion chamber.

Task B16 Inspect, adjust/replace cam followers.

Tappets, lifters, and CAM followers describe components that usually are positioned to directly ride or at least be actuated by, a cam profile. The term tappet has a broader definition and is sometimes used to describe what is more often referred to as a rocker lever. Because North American engine and fuel system original equipment manufacturers (OEMs) use all three terms, the technician should be familiar with them: in this text, the term tappet will not be generally used to describe a rocker and the term follower will be generally used to describe a component that rides the cam profile, except when describing specific OEM systems where an alternative term is preferred. The function of cam followers is to reduce friction and evenly distribute the force imparted from the cam's profile to the train it is responsible for actuating. Diesel engines using cylinder block mounted camshafts use two categories of follower while those using overhead camshafts use either direct actuated rockers or roller type cam followers.

Hydraulic lifters should be inspected during any head maintenance for:

- Hydraulic lifter body for scuffing and scoring. If the lifter body wall is worn or damaged, the mating bore in the block should also be checked.
- Check the fit of each hydraulic lifter in its mating bore in the block. If the clearance is excessive, try a new lifter.
- The hydraulic lifter foot must be smooth and slightly convex. If worn, pitted, or damaged, the mating camshaft lobe should also be checked.

Task B17 Adjust valve clearance(s).

When valves are properly adjusted, there should be clearance between the pallet and of the rocker arm and the top of the valve stem. Valve lash is required because as the moving parts wear they expand, and if clearance were not factored somewhere in the valve the valves would remain open by the time an engine reached operating temperature. Actual valve lash valves will depend on factors such as the length of the push tubes and the materials used in valve manufacture. Exhaust valves are subject to more heat and as a consequence, OEMs require a valve lash setting for exhaust valves that is usually greater than the intake valve lash setting.

Maladjusted Valves

A loose valve adjustment will retard valve opening and advance valve closing, decreasing the cylinder breathing time. Actuating cam geometry is designed to provide some "forgiveness" to the train at valve opening and valve closure to reduce the shock loading the train is subject to. When valves are set loose, the valve train is loaded at a point on the cam ramp beyond the intended point. The same occurs at valve closure when the valve is seated. High valve opening and closing velocities subject the valve and its seat to hammering which can result in cracking, failure at the head to stem fillet, and scuffing to the cam and its follower.

Valve Adjustment Procedure

The following steps outline the valve adjustment procedure on a typical 4 stroke cycle, in-line 6 cylinder diesel engine. Valves should always be adjusted using the OEMs specifications and procedure.

Caution: Shortcutting the engine OEM recommended valve (and injector) setting can result in engine damage unless the technician knows the engine well. Cam profiles are not always symmetrical and some engines may have camshaft profiles designed with ramps between base circle and outer base circle for purposes such as actuating engine compression brakes. Similarly, a valve rocker that shows what appears to be excessive lash when not in its setting position is not necessarily defective.

6 cylinder engine firing order: 1-5-3-6-2-4 cylinder throw pairings: 1-6 @ TDC
5-2 @ 1200 BTDC 3-4 @ 1200 ATDC

Cylinder throw pairings are sometimes called companion cylinders. In other words, when #1 piston is at TDC completing its compression stroke, #6 piston (its companion) is also at TDC having just completed its exhaust stroke. If the engine is viewed from overhead with the rocker covers removed, engine position can be identified by observing the valves over pair of companion cylinders. For instance, when the engine calibration plate indicates that the pistons in cylinders #1 and #6 are approaching TDC and the valves over #6 are both closed (lash is evident), then the point at which the valves over #1 cylinder rock (exhaust closing, intake opening) at valve overlap will indicate that #1 is at TDC having completed its exhaust stroke and #6 at TDC having completed its compression stroke. This method of orienting engine location is commonly used for valve adjustment.

Adjustment

1. Locate the valve lash dimensions. The lash specification for the exhaust valve(s) will usually (but not always) be greater than that for the inlet valve.

2. The valves on current diesel engines should usually be set under static conditions and with the engine coolant 1000°F (370°C) or less. Locate the engine timing indicator and the cylinder calibration indexes, 1,200 apart: depending on the engine, this may be located on a harmonic balancer, any pulley driven at engine speed, or the flywheel.

3. Ensure that the engine is prevented from starting by mechanically or electrically no-fuelling the engine. The engine will have to be barred in its normal direction of rotation through two revolutions during the valve setting procedure requiring the engine to be no-fuelled to avoid an unwanted startup.

4. If the engine is equipped with valve bridges or yokes that require adjustment, this procedure should be performed prior to the valve adjustment. To adjust a valve yoke, back off the rocker arm then loosen the yoke adjusting screw locknut and back off the yoke adjusting screw. Using finger pressure on the rocker arm (or yoke), load the pallet end (opposite to the adjusting screw) of the yoke to contact the valve: next, screw the yoke adjusting screw CW until it bottoms on its valve stem. Turn an additional one flat of a nut (1/6 of a turn), then lock to position with the locknut.

Caution: When loosening and tightening the valve yoke adjusting screw locknut, the guide on the cylinder head is vulnerable to bending. Most OEMs recommend that the yoke be removed from the guide and placed in a vice to back off and final torque the adjusting screw locknut.

5. To verify that the yoke is properly adjusted, insert two similarly sized thickness gauges of .010 in. or less between each valve stem and the yoke. Load the yoke with finger pressure on the rocker arm and simultaneously withdraw both thickness gauges: they should produce equal drag as they are withdrawn. If the yokes are to be adjusted they can be adjusted in sequence as each valve is adjusted. In some engines, valve yokes can only be adjusted with the rocker assemblies removed as it is impossible to access the yoke to verify the adjustment otherwise.

6. If the instructions in the OEM literature indicate that valves must be adjusted in a specific engine location, ensure that this is observed: the arms that actuate the valves may only have a small percentage of base circle. Setting valves requires the last dimension between the rocker arm and the valve stem on the rocker arm and the valve yoke to be defined. When performing this procedure, the valve adjusting screw locknut should be backed off and the adjusting screw backed out. Insert the specified size of thickness gauge between the rocker and the valve stem/yoke. Release the thickness gauge. Then turn the adjusting screw CW until it bottoms: turn an additional 1/2 flat of a nut (1/12 turn). Hold adjusting screw with a screwdriver and with a wrench lock into

position. Now for the first time since inserting the thickness gauge, handle it once again: withdraw the thickness gauge. A light drag indicates that the valve is properly set. If the valve lash setting is either too loose or too tight, repeat the setting procedure. Do not set valves too tight. Set all the valves in cylinder firing order sequence rotating the engine 1,200 between setting. It is preferable to begin at #1 cylinder and proceed through the engine in firing order sequence.

Task B18 Inspect, measure, and replace/reinstall overhead camshaft and bearings; measure and adjust end play and backlash.

A careful inspection of the camshaft will ensure that proper valve and injector timing can be obtained. Because the camshaft is constantly supplied with oil, cleaning should be as simple as rinsing with dip tank solution and blowing out with compressed air. Oil passages may require cleaning with a wire brush. Discard any camshaft that does not pass a visual inspection. If the camshaft passes the visual part of the inspection, take measurements and compare with manufacturer's specifications. The cam lobes must be inspected for signs of pitting, scoring, or flat spots. Use the manufacturer's procedures for taking lobe measurements. The heel-to-toe measurement is equal to the diameter of the circular portion of the lobe and the maximum amount of lift created by the rise at the top of the lobe. Typically, the cam follower will ride in the center of lobe travel causing a groove to form. This groove should not be greater than 0.003 in. (0.076 mm) deep. If either the top of the lobe or the circular portion is worn (grooved), the amount of lift will be reduced. If an intake or exhaust cam lobe fails to produce enough lift, the valve it controls may not open enough to supply adequate air for combustion to occur. If the lobe controls an injector, the injector may not fire. Visually inspect bearing journals for bluing, scoring, or wear. Measure the diameter of journals with an outside micrometer, and check for out-of-round condition by checking around the journal in several places. Inspect the shaft and gear keyway for cracks or distortion. Ensure the keyway in the gear and shaft matches the woodruff key. Any movement between these three can randomly change the timing of valves and injectors. A shift of as little as 0.004 in. (0.1 mm) in the keyway could equate to almost one degree in timing shift.

C. Engine Block Diagnosis and Repair (8 Questions)

Task C1 Remove, inspect, service, and install pans, covers, vents, gaskets, seals, and wear rings.

Oil pans are usually located in the airflow under the frame rails and are vulnerable to damage from objects on the road from rocks on rough terrain to small animals on the highway. Most highway diesel engines have oil pans that can be removed from the engine while it is in chassis. It is advisable to drain the oil sump prior to removing it. Oil pans that seal to the engine block using gaskets can be difficult to remove especially where adhesives have been used to ensure a seal. A 4 lb. rubber mallet may facilitate removal: avoid driving screwdrivers between the oil pan and its block mating flange because the result will be damage to the oil pan flange and the cylinder block mating flange. Oil pans that use rubber isolator seals tend not to present removal problems. After removal the oil pan should be cleaned of gasket residues and washed with a pressure washer. Inspect the oil pan mating flanges, check for cracks, and test the drain plug threads. Cast aluminum oil pans that fasten both to the cylinder block and to the flywheel housing are prone to stress cracking in the rear: with these oil pans always meticulously observe torque sequences and values. Cast aluminum oil pans can be successfully repair welded without distortion using the TIG (tungsten inert gas) process providing the cracks are not too large and the oil saturated area of the crack is ground clean. It is more difficult to execute a lasting repair weld on a stamped steel oil pan due to the distortion that results: steel oil pans may rust through in which case they should be replaced rather than welded. Whenever an oil pan has indications of porosity resulting from corrosion, it should always be replaced. The technician should always be aware of the fact that a failure of the oil sump or its drain plug on the road usually results in the loss of the engine.

Task C2 Disassemble, clean, and inspect engine block for cracks; check mating surfaces for damage or warpage; check condition of passages, core, and gallery plugs; inspect threaded holes, studs, dowel pins, and bolts for serviceability; service/replace as needed.

The block counterbore and packing ring area must be completely cleaned of all rust, scale, and grease. Most scale and/or rust can be removed using a piece of crocus cloth or 100 to 120 grit emery paper or wet-dry sandpaper. Inspect the counterbore closely for cracks, and ensure that no rough or eroded areas exist to ruin a sleeve or cut its O-ring. Damage of this nature may require re-sleeving the counterbore by an experienced mechanic with special tools. If the counterbore is in good shape, use a dial indicator mounted on a fixture to measure the counterbore top depth. A depth micrometer can be used in place of the dial indicator if firm contact is maintained with the block surface.

Check for deck warpage using a straightedge and thickness gauge: a typical maximum specification is going to be in the region of .1 mm or .004 in. Check the main bearing bore and alignment. Check the engine service history to ensure that the engine has not previously line bored. A master bar is used to check alignment for a specific engine series; check that the correct bar has been selected for the engine being tested and then clamp to position by torquing down the main caps minus the main bearings. The master bar should rotate in the cylinder block main bearing line bore without binding. Should it bind, the cylinder block should be line bored.

Check the cam bore dimensions and install the cam bushings with the correct cam bushing installation equipment: use drivers with great care as the bushings may easily be damaged and ensure that the oil holes are lined up prior to driving the bushing home.

Install gallery and expansion plugs are often interference fitted and sealed with silicone, thread sealants and hydraulic dope. Electromagnetic flux test the block for cracks at each out of chassis overhaul.

Task C3 Pressure test engine block for coolant leakage; service as needed.

After the cylinder block has been cleaned, it must be pressure tested for cracks or leaks by either one of two methods. One method can be used when a large enough water tank is available and the cylinder block is completely stripped of all parts:

- Seal off the water inlet and outlet holes airtight. This can be done by using steel plates and suitable rubber gaskets held in place by bolts. Drill and tap one cover plate to provide a connection for an airline.
- Immerse the block for twenty minutes in a tank of water heated to 180–200°F (82–93°C).
- Apply 40 psi (276 kPa) air pressure to the water jacket and observe the water in the tank for bubbles which indicate the presence of cracks or leaks in the block. A cracked cylinder block must be replaced.
- After the pressure test is completed, remove the block from the water tank. Then remove the plates and gaskets and dry the block with compressed air.
- Another method may be used when a large water tank is unavailable, or when it is desired to check the block for cracks without removing the engine from the vehicle. However, it is necessary to remove the cylinder heads, blower, oil cooler, air box cover, and oil pan.
- Attach sealing plates and gaskets as indicated in the first method. However, before attaching the last sealing plate, fill the water jacket with a mixture of water and one gallon of antifreeze. The antifreeze will penetrate small cracks and its color will aid in detecting their presence.
- Install the remaining sealing plate and tighten it securely.

- Apply 40 psi (276 kPa) air pressure to the water jacket and maintain this pressure for at least two hours to give the water and antifreeze mixture ample time to work its way through any cracks which may exist.
- At the end of the test period, examine the cylinder bores, air box, oil passages, crankcase, and exterior of the block for presence of the water and antifreeze mixture, which will indicate the presence of cracks. A cracked cylinder block must be replaced.

Task C4 Inspect cylinder sleeve counterbore and lower bore; check bore distortion; determine needed service.

Check the cylinder sleeve counterbore for the correct depth and circumference. Counterbore depth should typically not vary by more than .025 mm (.001 in.). Counterbore depth can be subtracted from the sleeve flange dimension to calculate sleeve protrusion. Recut and shim the counterbore using the OEM recommended tools and specifications.

Task C5 Inspect and measure cylinder walls or liners for wear and damage; determine needed service.

Before reinstalling, old, wet cylinder sleeves should be cleaned by immersion or glass bead blasting and closely inspected. Sleeves that fail the inspection should be discarded to ensure maximum overhaul life. The inspection should include checking for tapers using an inside micrometer, snap gauge, or cylinder gauge. An out-of-round condition can be detected by measuring the diameter at one point inside the cylinder and rotating the inside micrometer around the cylinder 180 degrees. To check liner to block height, first install and leave a liner clamping tool in place, holding the liner down. Next, measure the liner protrusion with a sled gauge. The liner protrusion should be in the .000 to .003 in. with no more than a .002 in. variance between adjacent liners.

Task C6 Deglaze and clean cylinder walls or liners.

After cleaning and pressure testing, inspect the cylinder block. Since most of the engine cooling is accomplished by heat transfer through the cylinder liners to the water jacket, a good liner to block contact must exist when the engine is operating. Whenever the cylinder liners are removed from an engine, the block bores must be inspected. Before attempting to check the block bores, hone them throughout their entire length until about 75 percent of the area above the ports has been "cleaned up."

Two basic methods of liner reconditioning are used: glaze busting and honing. Glaze busting: When checked to be within serviceability specs, the liner should be deglazed. Deglazing involves the least amount of material removal. A power driven flex hone or rigid hone with 200–250 grit stones. The best type of glaze buster is the flex hone, typically a conical (Xmas tree) or cylindrically shaped shaft with flexible branches with carbon/abrasive balls. The objective of glazebusting is to machine away the cylinder ridge above the ring belt travel and re-establish the crosshatch. The drill should be set at 120–180 rpm and used in rhythmic reciprocating thrusts: short sequences stopping frequently to inspect the finish produces the best results. You should observe a 60 degree crossover angle, 15–20 microinch crosshatch.

Honing: Use a hone in which the cutting radius of the stones can be set in a fixed position to remove irregularities in the bore rather than following the irregularities as with a spring loaded hone. Clean the stones frequently with a wire brush to prevent stone loading. Follow the hone manufacturer instructions regarding the use of oil or kerosene on the stones. Do not use such cutting agents with a dry hone.

Insert the hone in the bore and adjust the stones snugly to the narrowest section. When correctly adjusted, the hone will not shake in the bore, but will drag freely up and down the bore when the hone is not running.

Start the hone and "feel out" the bore for high spots which will cause an increased drag on the stones. Move the hone up and down the bore with short overlapping strokes about 1 inch long. Concentrate on the high spots in the first cut. As these are removed, the drag on the hone will become lighter and smoother. Do not hone as long at the air inlet port area as in the rest of the bore because this area, as a rule, cuts away more rapidly. Feed lightly to avoid an excessive increase in the bore diameter. Some stones cut rapidly even under low tension.

When the bore is clean, remove the hone, inspect the stones, and measure the bore. Determine which spots must be honed most. Moving the hone from the top to the bottom of the bore will not correct an out-of-round condition. To remain in one spot too long will cause the bore to become irregular. Where and how much to hone can be judged by feel. A heavy cut in a distorted bore produces a steady drag on the hone and makes it difficult to feel the high spots. Therefore, use a light cut with frequent stone adjustments. Overhead boring tools can be set to produce the required stroke rate for the specified crosshatch but if using a hand-held tool, remember that a few short strokes with a moderate radial load tend to produce a better crosshatch pattern than many strokes with a light radial load. You should observe a 60 degree crossover angle pattern, 15–20 microinch crosshatch. It should be clearly visible by eye. Wash the cylinder block thoroughly with soap and water after the honing operation is completed.

Task C7 Replace cylinder liners and seals; check and adjust liner height.

Selective fitting of liners to block bores is good practice whether or not the block bore has been machined. To selective fit a set of dry liners to a block, measure the inside diameter of each block bore across the north-south and east-west faces and grade in order of size. Next, get the set of new liners and measure the outside diameter of each: once again grade in order of size. Ensure that every measurement falls within the OEM specifications. Then fit the liner with the largest OD to the block bore with the largest ID. Loose fits of dry liners around .035 mm (.0015 in.) are common.

To pull wet-type sleeves from the cylinder block:

- Select the correct adapter plate that will fit the sleeve.
- Make sure the plate fits snugly in the sleeve (to prevent cocking) and that the outside diameter of the puller plate is not larger than the sleeve outside diameter. An adapter plate larger than the sleeve may damage the block.
- Attach the adapter plate to the through bolt.
- If the adapter plate is the type that can be installed from the top of the sleeve, it will have a cutaway or milled area on each side. This, along with the swivel on the bottom of the through bolt, allows the plate to be tipped slightly and inserted from the top, eliminating the need to install the adapter plate in the bottom of the sleeve and then inserting the through bolt and attaching a nut.
- After you install the adapter plate and through bolt in the sleeve, hold the through bolt and adapter plate firmly in the sleeve with one hand and install the support bracket with the other hand.
- Screw the through bolt nut down on the through bolt until it contacts the support bracket. This will hold the adapter plate and through bolt snugly in place.
- Before tightening the sleeve puller nut, make sure the sleeve puller supports or legs are positioned on a solid part of the block. Tighten the nut with a ratchet; the sleeve should start to move upwards. If it does not and the puller nut becomes hard to turn, stop and recheck your puller installation before proceeding.
- On wet-type sleeves, after the sleeve has been pulled from the block far enough to clear the O-rings, tip or swivel the sleeve puller adapter plate and remove the sleeve puller assembly.
- The sleeve can now be lifted out by hand.
- Engines with tight fitting dry-type sleeves may require a special hydraulic puller.

Task C8 Inspect in-block camshaft bearings for wear and damage; replace as necessary.

The camshaft is supported by pressure lubricated, friction bearings at main journals. The camshaft is subjected to loading whenever a cam is actuating the train that rides its profile. This loading can be considerable especially when engine compression brake hydraulics and fuel injection pumping apparatus are actuated by the camshaft. Cam bushings are normally routinely replaced during engine overhaul, but if they are to be reused they must be measured with a dial bore gage or telescoping gage and micrometer to ensure that they are within the OEM reuse parameters. Interference fit, cylinder block located, camshaft bushings are removed sequentially starting usually from the front to the back of the cylinder block using a correctly sized bushing driver (mandrel) and slide hammer. Cam bearing split shells are retained either cam cap crush (overhead camshafts) or lock rings (DDC 53, 71, and 92 Series). Interference fit cam bushings are installed using the same tools used to remove them. Care should be taken to properly align the oil holes: when installing bushings to a cylinder block in chassis where access and visibility are restricted, painting with whiteout (type correction fluid) the oil hole location on the bushing rim may help align the bushing before driving it. Ensure that the correct bushing is driven into each bore: bushings and their support bores vary dimensionally and new bushings rarely survive being driven into a bore and removal. Bearing clearance should be measured after installation. Bearing clearance is the camshaft bushing dimension measured with a dial bore gage or telescoping gage and micrometer minus the camshaft journal dimension measured with a micrometer. Prolonged use of certain types of engine compression brake can be especially hard on camshaft bearings.

Task C9 Inspect, measure, and replace/reinstall in-block camshaft; measure/adjust end play.

A careful inspection of the camshaft will ensure that proper valve and injector timing can be obtained. Because the camshaft is constantly supplied with oil, cleaning should be as simple as rinsing with dip tank solution and blowing out with compressed air. Oil passages may require cleaning with a wire brush. Discard any camshaft that does not pass a visual inspection. If the camshaft passes the visual part of the inspection, take measurements and compare with manufacturer's specifications. The cam lobes must be inspected for signs of pitting, scoring, or flat spots. Use the manufacturer's procedures for taking lobe measurements. The heel-to-toe measurement is equal to the diameter of the circular portion of the lobe and the maximum amount of lift created by the rise at the top of the lobe. Typically, the cam follower will ride in the center of lobe travel causing a groove to form. This groove should not be greater than 0.003 in. (0.076 mm) deep. If either the top of the lobe or the circular portion is worn (grooved), the amount of lift will be reduced. If an intake or exhaust cam lobe fails to produce enough lift, the valve it controls may not open enough to supply adequate air for combustion to occur. If the lobe controls an injector, the injector may not fire. Visually inspect bearing journals for bluing, scoring, or wear. Measure the diameter of journals with an outside micrometer, and check for out-of-round condition by checking around the journal in several places. Inspect the shaft and gear keyway for cracks or distortion. Ensure the keyway in the gear and shaft matches the woodruff key. Any movement between these three can randomly change the timing of valves and injectors. A shift of as little as 0.004 in. (0.1 mm) in the keyway could equate to almost one degree in timing shift.

Camshaft end play is defined either by free or captured thrust washers/plates. Thrust loads are not normally excessive unless the camshaft is driven by a helical toothed gear, in which instance there is more likely to be wear at the thrust faces. End play is best measured with a dial indicator: the camshaft should be gently levered longitudinally rearward then foreword and the travel measured and checked to specifications.

Task C10 Clean and inspect crankshaft for surface cracks and journal damage; check condition of oil passages; check passage plugs; measure journal diameters; determine needed service.

An inspection of the crankshaft begins with the connecting rod and main bearing shells. The bearing shells will usually show signs of damage before the crank. If the bearings indicate damage, inspect the crankshaft journal associated with that shell before it is cleaned. Toughly clean the crank and visually inspect for obvious damage. This includes cracked or worn front hub key slots, the main bearing journal for scoring of bluing, dowel pins and holes for cracks and wearing, oil supply holes for cracks, and the shaft for seal grooving. In addition, a careful inspection of the crank is performed by taking measurements. An out-of-round journal can be detected by using an outside micrometer to measure in at least two places around the journal's diameter. Generally, this measurement should not exceed 0.002 in. (0.051 mm) and a medium out-of-round journal is not acceptable for use in a maximum out-of-round bore. To find a tapered journal, measure the diameter nearest each side and the middle of the journal. Examples of these measurements are no more than 0.0015 in. (0.381 mm) between any two, no more than 0.0005 in. (0.013 mm) between tapers in the journal and bearing. Check thrust surfaces visually for scoring, nicks, and gouges. Measure clearance with an inside micrometer. Visually inspect for obvious cracks or breaks, and use either magnetic particle or spray penetration methods of detection to find smaller, less obvious ones.

Crankshaft Failures

Bending failures; abnormal bending stresses occur when:

- Main bearing bores are misaligned. Any kind of cylinder block irregularity will cause abnormal stress over the affected crankshaft area.
- Main bearings fail or become irregularly worn.
- Main caps are broken or loose.
- Standard specification main bearing shells are installed where an oversize is required.
- The flywheel housing is eccentrically (relative to the crankshaft) positioned on the cylinder block. A possible outcome of failure to indicate a flywheel housing on reinstallation.
- Crankshaft is not properly supported either out of the engine prior to installation or while replacing main bearings in-chassis: the latter practice is more likely to damage the block line bore but may deform the crankshaft as well.
- Bending failures tend to initiate at the main journal fillet and extend through to throw journal fillet at 90 degrees to the crankshaft axis.

Torsional Failures

Excessive torsional stress can result in fractures occurring from an initiation point in a journal oil hole extending through the fillet at a 450° angle or circumferential severing through a fillet. In an in-line 6 cylinder engine, the #5 and 6 journal oil holes tend to be vulnerable to torsional failure when the crankshaft is subjected to high torsional loads.

Causes of crankshaft torsional failures:

- Loose, damaged, or defective harmonic damper or flywheel assembly.
- Unbalanced engine driven components; fan pulleys and couplings, fan assembly, idler components, compressors, and PTOs.
- Engine overspeed. Even fractional engine overspeeding will subject the crankshaft to torsional stresses that it was not engineered for that may initiate a failure. It should be remembered when performing failure analysis that the event that caused the failure and the actual failure may be separated by a considerable time span.
- Unbalanced cylinder loading. A dead cylinder or fuel injection malfunction of over or under fueling a cylinder(s) can result in a torsional failure.
- Defective engine mounts. This can produce a "shock load" effect on the power train.

Task C11 Check main bearing bore alignment and cap fit; determine needed service.

Inspection of the main bearing bore alignment and out-of-roundness can prevent possible damage to the crankshaft or premature bearing failure. If the bearing bore is out of alignment uneven pressure is applied to each of the bearings, causing the crank to flex as it is rotated. This flexing and its effects can also be created by an operator who "hot rods" or when the engine becomes lugged down. In this case, the engine will heat unevenly, resulting in uneven expansion of the journal. Verifying the alignment using a master bar may prove the latter. This flexing can cause the crank to fracture and will result in bearings with unusual wear patterns. Use an inside micrometer to check the bearing bore for correct diameter and an out-of-round condition. If the bearing bore is too large, bearing shells will not seat properly against the bore. This may cause inadequate heat transfer from the shells to the bore, turning, or movement of shells in the bore, and misalignment of the crank when load is applied. An out-of-round bore will cause uneven torque on the crank, which could lead to fatigue, fracturing, and failure or breakage of the crank. The bearing shells will not be lubricated evenly and will wear more on one side, while the distorted shell will create fractures in the bearing surface.

Task C12 Inspect and replace main bearings; check bearing clearances; check and adjust crankshaft end play.

Correct installation of bearing shells is essential to any successful overhaul. About 13 percent of premature bearing failures are caused by improper assembly of bearings. After reconditioning the block and selecting proper sized bearings, carefully install the top half of each bearing into the cylinder block. Ensure the locating lugs (indented notch in each shell) are fitted into the matching slot in the cylinder block bore. This type of error can cause movement of bearings leading to metal in oil and finally a spun bearing causing oil starvation and seizing of the crank. Sight through the oil passage in the shell to ensure that it is aligned to the oil passage in the bore. Failure to align the oil passages will result in oil starvation, bearing fatigue, scoring, metal in oil, and finally, other bearing failure. Carefully install the rear main seal (split type), crank in the block, and align marks on the crank timing gear and cam gears. Use a Plastigage® to check all main bearing clearances. If clearances are not within specification, check for improperly sized bearings, dirt, or metal under shells, or misalignment of shells. If clearances are correct, remove the Plastigage, lubricate bearings, reinstall bearing caps, and torque to specifications. Rotate the crank by hand to check for binding. If binding occurs, loosen all caps and tighten individually to determine which bearing is at fault. Check and adjust crankshaft end play.

Spun Bearing(s)/Bearing Seizure

A lubrication related failure is caused by insufficient or complete absence of oil in one or all the crank journals. The bearing is subjected to high friction loads and surface welds itself to the affected crank journal. This may result in a spun bearing in which the bearing friction welds itself to the journal and rotates with the journal in the bore or alternatively continue to scuff the journal to destruction. When a crankshaft fractures as a result of bearing seizure, the surface of the journal is destroyed by excessive heat and it fails because it is unable to sustain the torsional loading.

Causes of spun bearings and bearing seizure:

- Misaligned bearing shell oil hole.
- Improper bearing to journal clearance. Excessive clearance will result in excessive bearing oil throw-off which starves journals furthest from the supply of oil. Insufficient bearing clearance can be caused by over torquing, use of oversize bearing shells where a standard specification is required, and cylinder block line bore irregularities.
- Sludged lubricating oil causing restrictions in oil passages.
- Contaminated engine oil. Fuel or coolant in lube oil will destroy its lubricity.

Etched main bearings are caused by the chemical action of contaminated engine lubricant. Chemical contamination of engine oil by fuel, coolant, or sulfur compounds can result in high acidity levels which can corrode all metals. However, the condition is usually first noticed in engine main bearings. It may result from extending oil change intervals beyond that recommended. Etching appears initially as uneven, erosion pock marks, or channels.

Task C13 Inspect, reinstall, and time the drive geartrain. This includes checking timing sensors, gear wear, and backlash of crankshaft, camshaft, auxiliary, drive, and idler gears; service shafts, bushings, and bearings.

The timing geartrain, to include all gears that drive the camshaft (including the crankshaft gear), must be inspected for damage and wear. The teeth to the gears must be in good shape to maintain the timing of intake and exhaust valves and on some engines, the injectors. Excess tooth wear can seriously affect engine performance and long term operation and may cause damage to the piston, cylinder, and crank. Carefully inspect for a slight roll or lip on each gear tooth, in addition to looking for chipping, pitting, and burring. If these are present on any gear replace it and examine the mating gears for similar damage. Try to determine the cause of damage. If removal is necessary, generally a press or puller will be required. Ensure the gear is properly supported on the center of the hub next to the shaft to prevent cracking or breaking and protect the end of the camshaft. For removal of intermediate and crankshaft gears, refer to the specific sections of the manufacturer's manuals. Installation of these gears is very critical as it will affect the timing of the engine. The camshaft is generally keyed the same between the shaft and gear, although some manufacturers use offset keys to shift the keyway and thereby advance the timing. To reduce emissions, modern engines are incorporating electronic camshaft advance mechanisms called variable cam phasing.

Task C14 Clean, inspect, measure, or replace pistons, pins, and retainers.

After removing each piston, carefully inspect it and its matching cylinder for possible signs of damage and try to determine probable cause before cleaning. Check for scoring in the skirt of the piston and compare it to the matching side(s) of the cylinder. This could have been caused by engine overheating, excessive fuel setting, improper piston clearance, insufficient lubrication, or an improper injector or nozzle. A crack in the skirt or lands may be accompanied by a bluing or gouges in the skirt and/or cylinder matching with the damage. This may have resulted from excessive use of starting fluid or introduction of starter fluid at operating temperatures, excessive piston clearance, or foreign objects in the cylinder like a dropped valve. A nick in the bottom of the skirt and connecting rod, or the cylinder and connecting rod may be caused by improperly installed connecting rods. Uneven wear in the skirt matching a bluing or scuffing on the cylinder could be caused by normal operation at low temperatures, dirty lubricating oil, too little piston clearance, or dirty intake air. The affects of stuck or broken rings may appear on the cylinder as scrapes and gouges, scratches and carbon build-up, or glazing. The leading causes are overheating, insufficient lubrication, and excessive fuel settings. Worn piston pin bores can also be caused by normal wear, insufficient lubrication, or dirty lubricating oil. Under normal conditions, during an overhaul the piston and rings are replaced.

Piston fit problems:

- Excessive piston to bore clearance—results in piston knocking against cylinder wall—especially noticeable when engine is cold.
- Too little piston clearance—causes piston scoring and scuffing (localized welding)—the film of lube oil on the cylinder wall is scraped off.

Piston and Rings Inspection and Failure Analysis

When clamping a piston/connecting rod assembly in a vice, use brass jaws or a generous wrapping of rags around the connecting rod: the slightest nick or abrasion may cause a stress point from which a failure could develop. When attempting to diagnose the cause of a piston failure, the following may be used as a very general guideline:

Skirt scoring—causes: overheating, overfueling, improper piston clearance, insufficient lubrication, improper injector nozzle.

Cracked skirt—causes: excessive use of ether, excessive piston to bore clearance, foreign objects in cylinder.

Uneven skirt wear—causes: dirt in lube oil, abrasives in air charge, too little piston to bore clearance.

Broken ring lands—causes: excessive use of ether, excessive piston to bore clearance, foreign objects in cylinder.

Worn piston pin bosses—causes: old age (normal wear), contaminated lube oil, insufficient lubrication.

Burned or eroded center crown—causes: plugged nozzle orifice, injector dribble, cold loading of engine, retarded injection timing, water/coolant leakage into cylinder.

Fuel injection/ignition timing related failures of pistons:

Advanced engine timing causes: excessive combustion pressures and temperatures. It may result in torching or blowing out the lower ring lands in severe cases. In less severe cases, erosion and burning of the top ring land or headland area result, evidenced by pitting.

The latter condition is more common and is often the result of abuse by technicians occasioned by the belief it will increase engine power. It often does, but at a heavy cost, a significant reduction in engine longevity. Total failure will usually occur in one cylinder with the evidence in the remaining cylinders.

Retarded Engine Timing

This will cause excessive cylinder temperatures and burning/erosion damage through the central crown area of the piston. An indicator of retarded injection timing is piston crown scorching under the injector nozzle orifice. The condition is seen much less frequently than advanced injection timing because it is seldom intentional and usually the result of technician error or component failure.

Torched Pistons

This term is used to describe a piston or set of pistons that have been overheated to such an extent that melt-down has occurred. Diagnosing a torched piston in isolation from the engine from which it was removed may be a game of guesswork rather than sound failure analysis practice. However, the cause of the condition must be unmistakably diagnosed before the engine is reassembled or a recurrence of the failure will result. If only one of a set of pistons is affected, the diagnosis may be simpler but it is always important to remember that piston torching may be related to lubrication, cooling system, or fuel injection causes.

Task C15 Measure piston-to-cylinder wall clearance.

Use a micrometer to measure the piston diameter at right angles to the piston pin bore and one inch below the bottom edge of the lowest ring groove. Compare these measurements to the bore diameter of the parent bore. The difference between these two valves becomes the piston-to-cylinder wall clearance. The average running clearance is .006 in. (.152 mm), but not less than .002 in. (.050 mm). Insufficient clearance will cause premature piston or cylinder/liner failure. You measure cam ground pistons at right angles (90 degrees) to the piston pin. The running clearance of all pistons can be measured using a spring scale with a feeler gauge attached to its end. Insert the .006 in. (.152 mm) feeler gauge in the cylinder (liner). Lubricate the piston with oil and install it (with no rings) bottom up, with the feeler gauge between the cylinder ID and the piston. Position the piston about 2 in. (50 mm) below the fire deck with the piston pin bore in line with the crankshaft. When the spring scale indicates the specified force in pounds as the feeler gauge is being withdrawn the running clearance is correct.

Task C16 Check ring-to-groove clearance and end gaps; install rings on pistons.

Ensure the rings have the proper end gap before installing them on the piston. This measurement is taken while the rings are in the cylinder. Slide each ring into the cylinder level using a ringless piston, and measure the gap between ends of the ring. Remove rings from the cylinder and follow manufacturer's instructions for installing these specific rings on the piston. When installing the piston in the cylinder ensure the ring gaps are spaced 90 to 120 degrees apart. Ring gap specification is usually .003 in.–.004 in. per 1 in. of diameter. Ring side clearance is the installed clearance between the ring and the groove it is fitted to. The dimension is measured using thickness gauges.

Cold stuck/gummed rings—a condition in which the piston ring seizes in the ring groove when the engine is cold but frees and starts to seal the cylinder when the engine is at running temperature—caused by carbon/ sludge buildup in the ring groove. The condition results in high crankcase pressure during engine warmup and fuel contaminated engine lubricating oil.

Hot stuck rings—causes: crystalized carbon buildup resulting from the use of contaminated or improper fuel or the failure of oil control rings. The ring is usually captured stationary in the groove and the contact face blackened by cylinder blowby gas.

Scuffed rings—causes: usually related to an improper ring to cylinder bore fit such as would occur with an out of specification ring end gap, piston to bore clearance, or ring to groove side clearance. May also be caused by cylinder wall lubrication failure. Scuffing is a localized weld condition, the heat for which is created by friction so metal to metal contact is required. It often results in the seizure of the piston in the liner bore.

Eroded/glazed rings—causes: usually an indication that poorly filtered or unfiltered intake air has been breathed into the engine cylinders. Engines used in construction and aggregate hauling applications often ingest enough ultra-small particles of abrasive silica through a properly functioning air cleaner system to glaze and erode rings and liners and shorten engine life. However, whenever this condition is identified, always test the engine intake system for leaks.

Broken rings—causes: improper installation or crystallized carbon buildup in a sector of the ring groove. A condition that is not often observed in the current generation of engines due to the fact that brittle cast irons are seldom used as a ring material any more.

Plugged oil control rings—causes: low engine operating temperatures or broken down, sludged engine lubricant. The result of this condition is usually an engine that burns engine oil because the oil that is applied to the liner walls by the piston and bearing throw-off is not controlled. The condition will cause coincidental problems such as compression ring gumming/sticking.

Task C17 Identify piston and bearing wear patterns that indicate connecting rod alignment or bearing bore problems; check bearing bore and bushing condition; replace as needed.

Uneven skirt wear—causes: dirt in lube oil, abrasives in air charge, too little piston to bore clearance.

Task C18 Assemble pistons and connecting rods and install in block; replace rod bearings and check clearances; check condition, position, and clearance of piston cooling jets (nozzles).

Installation of piston pins into the aluminum piston can be made easier by preheating the piston in 200°F (93°C) water. Slide the piston pin through holes in the piston and connecting rod and install retaining rings. Clamp in a soft jawed vise by the connecting rod to support the assembly while installing rings. Ensure the rod number is correct for the cylinder in which it is being placed, and that number is facing the cam on six cylinder engines or the outside on V8 Engines. Protect the rod bearing and journal by covering rod bolts with plastic caps or rubber hose, and carefully guide the rod onto the journal until the bearing shell seats. Verify that bearing clearances are within specification using a Plastigage and connecting rod and cap numbers match their cylinder. Recheck the torque of all cap bolts.

Task C19 Inspect, measure, and service/replace crankshaft vibration damper.

The vibration damper reduces the amplitude of vibration and add to the flywheel's mass in establishing rotary inertia. Its primary function is to reduce crankshaft torsional vibration. They consist of a damper drive or housing and inertia ring. The housing is coupled to the crankshaft and using springs, rubber, or viscous medium, drives the inertia ring. The objective is to drive the inertia ring at average crankshaft speed. Viscous type harmonic balancers have become almost universal in truck and bus diesels; the annular housing is hollow and bolted to the crankshaft. Within the hollow housing, the inertia ring is suspended in and driven by silicone gel. The shearing of the viscous fluid film between the drive ring and the inertia ring effect the damping action. Most OEMs recommended the replacement of the harmonic balancer at each major overhaul but this is seldom observed due to the expense and practice which has shown these components to frequently exceed OEM projected expectations. The consequences of not replacing the damper when scheduled are economic as they may result in a failed crankshaft. The shearing action of the silicone gel produces friction which is released as heat. This leads to eventual breakdown of the silicone gel, a result of prolonged service life or old age. Drive housing damage is another common reason for viscous damper failure and this is probably caused by careless service facility practice in most cases.

Rubber type vibration dampers are not often observed on todays heavy truck diesel engines suggesting that they are probably less effective. This type consists of a drive hub bolted to the crankshaft: a rubber ring is bonded both to the drive hub and the inertia ring. The rubber ring therefore acts both as the drive and the damping medium. The inherent elasticity of rubber enables it to function as a damping medium, but the internal friction generates heat which eventually hardens the rubber and renders it less effective and vulnerable to shear failures.

To ensure that the balance of the crankshaft drivetrain is maintained during assembly of the engine, mark alignment of all drivetrain parts before removal. Inspect the vibration damper, flywheel/flexplate, and clutch/torque converter for damage that may cause an unbalanced condition. If damage is detected there may also have been uneven wear on the main bearings and/or stress on the crankshaft. Inspect the flywheel for cracks, missing or damaged ring gear teeth, distortion, or oblong mounting holes, or a pilot bearing with a loose or improper fit.

Visually inspect the damper housing noting any dents or signs of warpage: evidence of either is reason to reject the component. Use the following process:

- Using a dial indicator, rotate the engine manually and check for damper housing radial and axial runout against the OEM specification. This is a low tolerance specification usually .005 in. (0.127 mm) or less.
- Check for indications of fluid leakage, initially with the damper in place. Trace evidence of leakage justifies replacement of the damper.
- Should the damper not be condemned using the above tests, remove it from the engine. By hand, shake the damper: any clunking or rattle is reason to replace it.
- Next, using a gear hotplate or component heating oven, heat the damper to its operating temperature, usually close to the engine operating temperature, around 90°C (180°F)—this may produce evidence of a leak.
- A final strategy is to mount the damper in a lathe using a suitable mandrel. It should be run up through the engine operating range and monitored using a balance sensor and strobe light.
- While on the crankshaft check for wobble using a dial indicator, and replace if out of manufacturer's specifications. Viscous dampers should be checked for nicks, cracks, or bulges. Cracks or bulges may indicate that internal fluid has ignited and

caused the case to expand. Rebalance the crankshaft drivetrain if any major part is replaced.

Task C20 Inspect, install, and align flywheel housing.

Checking Flywheel Runout

You mount a dial indicator with a magnetic base on the flywheel housing. Check the flywheel runout at the clutch contact face. The typical allowable runout is .001 in. per inch of clutch radius (.001 mm per millimeter of clutch radius). You measure the radius from the center of the flywheel to the outer edge of the clutch contact face. For example, a 14-inch (355 mm) clutch would have a 7-inch (177 mm) radius and allow a .007 inch (.177 mm) total indicator reading of the dial indicator. If the measured values fall out of these specifications, you should repeat the assembly process and remeasure until a cause can be determined.

Flywheel Housing Alignment, Bore Concentricity, and Bolting Flange Runout

Any time a flywheel is removed from the crankshaft and cylinder block, the technician must check the housing inner flange face concentricity with the crankshaft using a dial indicator. The maximum tolerance for the crankshaft to flywheel housing eccentricity is low, typically around .012 in. (.3 mm) total indicated reading (TIR) and the consequences of installing a flywheel housing that exceeds the allowable specification are severe. When flywheel eccentricity is present the drive axis is broken, a condition that can result is clutch, engine mount, transmission and engine failures. The flywheel housing-to-crankshaft concentricity is preserved by interference fit cylindrical or diamonds dowels. These dowels may be relied upon to properly realign the flywheel housing on the cylinder block at each reinstallation, but the specification is so critical it should be checked. The following procedure outlines a method of checking flywheel housing concentricity and a couple of typical strategies for correcting an out of specification condition.

1. Ensure that the engine is properly supported. Indicating flywheel housing may be performed with the engine in or out of the chassis. Locate the flywheel inner flange face to crankshaft concentricity TIR tolerance in the OEM service manual. Either mechanically or electrically ensure the engine fuel system will not function so the engine will not start.
2. Mount the flywheel housing to the engine cylinder block using the dowels to align the assembly and snug the flywheel fasteners at about half of the OEM-specified torque value.
3. Use fabricator's chalk to mark the flywheel flange face in four locations with the following rotational positions: A, B, C, D.
4. Next, thoroughly clean the inside face of the flywheel housing with emery cloth and fix a magnetic base dial indicator to any position on the crankshaft setting the probe to contact the flywheel housing inside face. Using an engine barring tool, rotate the engine in its normal direction of rotation until the indicator probe is positioned at any one of the chalk indexes. Now set the indicator at zero.
5. Bar the engine through a full rotation, recording the indicator reading at each of the indexes. The indicator should once again read zero when the revolution is complete. If the readings are A=0, B=minus .003 in., C=minus .005 in., and D=plus .004 in. the TIR would be the highest negative value (minus .005 in.) added to the highest positive value (.004 in.) giving a reading of .009 in. If the OEM TIR maximum specification was .012 in. this reading would be within it. If the flywheel housing-to-crankshaft concentricity is outside of the specification, it can be reset.

The following outlines a Caterpillar Engine procedure for an engine that aligns the flywheel housing using cylindrical locating dowels.

Begin by removing the dowels. Check the availability of oversize locating dowels with the parts department. Loosen the flywheel mounting capscrews until they are barely snug, permitting the flywheel housing to be moved fractionally when struck with a rubber mallet. Perform the flywheel housing concentricity sequence as outlined above. Adjust the position of the flywheel housing by tapping with the rubber mallet on the basis of the readings of the dial indicator. Repeat the procedure until the readings meet the required specification. When they do, fully torque the flywheel housing capscrews to specification. Next, the locating dowel holes in the cylinder block and flywheel housing will have to be reamed to an oversize specification, factoring in the interference fit requirement.

Drive the oversize dowels into the reamed holes.

Install the flywheel housing as follows:

- Thread aligning studs into the cylinder block to guide the housing into place.
- Support the housing and position it over the crankshaft.
- Install the housing bolts and torque them in the OEM tightening sequence.

Task C21 Inspect crankshaft flange and flywheel/flexplate mating surfaces for burrs; measure runouts; repair as needed.

Reconditioning and Inspecting Flywheels

Flywheels are commonly removed from engines for reasons such as clutch damage, leaking rear main seals, leaking cam plugs etc. and care should be taken both when inspecting and reinstalling the flywheel and the flywheel housing. Flywheels should be inspected for face warpage, heat checks, scoring, intermediate drive lug alignment and integrity, axial and radial runout using dial indicators, straightedges, and thickness gauges. Damaged flywheel faces may be machined using a flywheel resurfacing lathe to OEM tolerances: typical maximum machining tolerances range from 0.060 to 0.090 in. (1.50 mm –2.30 mm). It is important to note that when resurfacing pot type flywheel faces, the pot face must have the same amount of material ground away as the flywheel face: the consequence of machining the clutch face only is to have an inoperable clutch.

Task C22 Inspect flywheel/flexplate (including ring gear) for cracks, wear, and runout; determine needed repairs.

Inspect the flywheel for cracks, missing or damaged ring gear teeth, distortion, or oblong mounting holes, or a pilot bearing with a loose or improper fit. Inspect the vibration damper, flywheel/flexplate, and clutch/torque converter for damage that may cause an unbalanced condition. If damage is detected there may also have been uneven wear on the main bearings and/or stress on the crankshaft.

Ring Gear Replacement

Shrunk fit to the outer periphery of the flywheel is the ring gear, which is the means of transmitting cranking torque to the engine by the starter motor during startup. A worn or defective ring gear is removed from the flywheel by first removing the flywheel from the engine and then using an oxyacetylene torch to partially cut through the ring gear working from the outside. One uses heating to expand the ring gear so that the removal can be completed using a hammer and chisel.

To install a new ring gear, place the flywheel on a flat, level surface and check that the ring gear seating surface is free from dirt, nicks, and burrs. Ensure that the new ring gear is the correct one and if its teeth are chamfered on one side, that they will face the cranking motor pinion after installation. Next, the ring gear must be expanded using heat so that it can be shrunk to the flywheel. Most OEMs specify a specific heat value because ring gears are heat treated and overheating will damage the tempering and substantially reduce the hardness. A typical temperature specification would be around 200°C (400°F). Because of its size, the only practical method of heating a ring gear for installation is using a rosebud type (high gas flow), oxyacetylene heating tip. To ensure that the ring gear is heated evenly to the specified temperature and especially to ensure that it is not overheated, the use of a temperature indicating crayon such as Tempilstick is recommended. When the ring gear has been heated evenly to the correct temperature it will usually drop into position and almost instantly contract to the flywheel.

D. Lubrication and Cooling Systems Diagnosis and Repair (11 Questions)

Task D1 Check engine oil pressure, pressure gauge, and sending unit.

Oil pressure gauges are either electrically or mechanically operated and low oil pressure lights are electrically operated indicators. These devices must be considered any time incorrect or faulty oil pressure readings are obtained. Usually gauges are extremely reliable, but depending on which type is used, there are things that can go wrong. Mechanical gauges use the engine oil pressure routed through a tube to operate the meter movements. If the tube becomes punctured, chinked, or clogged the fluid will not be able to operate the meter movements. In this case, the tube must be replaced if it is punctured or chinked or blown clean if clogged. Common problems associated with electrical gauges are open or shorted wiring or circuitry, clogged passage to the sensor, or a defective sensor. The meter can become damaged if too much voltage is applied, but this usually is a symptom of an electrical wiring problem. Use a Volt/Ohm meter (VOM) to check the resistance of the sensor and continuity of the wiring. If the sensor is at fault replace it and check the oil passage for obstructions. If the wiring is at fault repair or replace it and determine the cause. Finally, if all else appears to be working well, replace the meter. Indicator lights use a method similar to the electrical gauge except they incorporate a pressure switch and a light bulb. If sludge deposits are found at the fault of a failure of oil pressure, sensing a more serious problem exists, and the engine may need an overhaul. Do not attempt to flush the engine by running lightweight oils or a mixture of diesel fuel and oil, as this may result in damage that is more serious to the engine from insufficient lubrication and suspended particles. If the engine oil pressure is actually low, ensure that the oil is of the proper grade as specified by the manufacturer, and that the by-pass valve is properly set before considering the engine for overhaul.

Task D2 Inspect, measure, repair/replace oil pump, drives, inlet pipes, and screens.

When removing and inspecting engine oil pumps refer to the actual manufacturer's manual for specific instructions and required tools. Some manufacturers, like Cummins Diesel, use external oil pumps that have oil lines routed out of the block and return near the by-pass valve. Most other manufacturers use internal oil pumps that can be accessed from the crankcase by dropping the oil pan with conventional tools. Remove the bolts that hold the oil pump to the engine and then carefully remove the oil pump. With external oil pumps, remove the cover from the pump body and inspect it for wear. Remove and inspect the idler and driven gears for pitting and their teeth for wear. You should check the diameter of the gears and driveshaft using a micrometer and compare with manufacturer's specifications. Check all mating surfaces and the pump body for damage and wear. After reinstalling the gears and drive shaft in the body, check the gear-to-cover clearance with a Plastigauge. If it fails to meet manufacturer's specifications, replace the worn parts. If the pump includes a by-pass valve, it must also be checked for wear and cleaned.

Task D3 Inspect, repair/replace oil pressure regulator valve(s), by-pass valve(s), and filters.

Oil filters trap particles that would otherwise cause damage to other critical components, like engine bearings. Regular replacement of filter elements and inspection of the housing and other component parts is necessary. Modern engines use a variety of materials from cast iron to diecast aluminum for the housing and steel or plastic for machined parts. For this reason, the technician is limited to a visual inspection of filter assembly parts. The visual inspection must be complete and thorough. If there is doubt, or a component is questionable, replace it. Cracks anywhere in the housing and nicks on machined surfaces can cause major engine failures, and as such must be detected and appropriate action taken to correct the problem. Some filter assemblies incorporate a by-pass valve in their housing. These valve assemblies must be removed and inspected for corrosion, wear, and other signs of damage during overhaul or any time the filter assembly contains metal particles. Canister filters should be cut open and the element material examined for content. Suspended metal particles indicate a potential for bearing failure. A milky gray sludge indicates water in the oil caused by a possible head or cylinder block leak, which may already have caused bearing damage. Ensure all gasket surfaces are straight and free of nicks, which could cause an improper seal of the assembly. On element type filter assemblies, ensure that housing grooves, center bolt threads, springs, metal gaskets, and spacers are in good condition. Inspect for distortions in the canister, cracks around canister bolt hole and clogged passageways in the bolt. Reinstall the by-pass valve according to manufacturer's installation procedures and set for proper pressure. Replace the element and all gaskets and supplied seals. When reassembling filter assemblies, apply a small amount of oil to all seals to ensure a tight seal. Prefilling of canister type filters is recommended to ensure adequate oil is present during startup to lubricate the bearings.

Task D4 Inspect, clean, test, reinstall/replace, and align oil cooler; test, reinstall/replace by-pass valve and oil thermostat; inspect and repair/replace lines and hoses.

Complete servicing of the oil cooler should be done as part of a major engine overhaul. In addition, the cooler failures are oil in the water, water in the oil pan, or milky gray sludge in the filter. In either of these cases, specific manufacturer's instructions will apply.

Task D5 Inspect and clean turbocharger lubrication system; repair/replace as needed.

To check the turbocharger lubrication begin by checking for signs of leaks. Inspect hoses and tubes for damage, such as frayed or cut hoses and tubes for kinks or nicks. Remove hoses and tubes and check for obstructions. Install the lines after replacing any damaged ones and install a tee in one of the connections. Attach an oil pressure gauge to the tee and tighten all connections. Start the engine and monitor the oil pressure gauge. Operate the engine over a range of operating speeds while monitoring the oil pressure. Stop the engine and ensure the pressure drops. All gauge readings must mimic the instrument panel oil pressure gauge.

Task D6 Inspect, reinstall/replace, and adjust drive belts.

To install drive belts:

- When two or more identical belts are used on the same pulley, all of the belts must be replaced at the same time.
- Make sure the distance between the pulley centers is as short as possible when you install the belts. Do not roll the belts over the pulley. Do not use a tool to pry the belts onto the pulley.
- The pulleys must not be out of alignment more than 1/16 in. (1.59 mm) for each 12 in. (30.5 mm) of distance between the pulley center.
- The belts must not touch the bottom of the pulley grooves. The belts must not protrude more than 3/32 in. (2.38 mm) above the outside diameter of the pulley.
- When identical belts are installed on a pulley, the protrusion of the belts must not vary more than 1/16 in. (1.59 mm).
- Make sure that the belts do not touch or hit against any part of the engine.

To adjust drive belts:

- Use the belt tension gauge to check the tension of the belts.
- Tighten the reading of 110 to 160 lbs. as indicated on the gauge.
- After the engine has been running for at least one hour, stop the engine and check the belt tension. If the tension is less than the value given in the previous step, adjust the belt to the correct value.

Task D7 Check coolant temperature, temperature sensor, temperature gauge, and sending unit.

Coolant temperature is used by modern diesels to monitor and control the engine operation as well as keeping the driver informed. Standard temperature gauges and indicator lights fall into one of two types: mechanical or electrical. Mechanical gauges use a liquid filled sensor attached to the hydraulic meter movement by a tube. This type gauge is self-contained and must be replaced as an assembly. Common reasons for failure of a mechanical gauge are corrosion on sensor, leaking sensor, pinched tube, and binding of meter movement of needle. Electrical gauges use a sensor that changes its resistance as temperature changes. This resistance is used to vary the current applied to the meter movements. Failures of an electrical gauge can be caused by defective sensors, an open or short in wiring, and open or bad grounds. While meter movements will fail, they are usually a symptom of other problems. Indicator lights typically use a sensor that acts like a switch to turn on a light circuit. Probable causes for failure are the same as an electrical gauge except the light bulb itself can fail without any indication. With the introduction of electronics into the operation of diesel engines, new problems have been added. Most sensors are not directly connected to their indicators. As such, the problem may be with a faulty module or circuit card. Some problems may be introduced by software programs or other sensors used as part of an overall calculated program.

Task D8 Inspect and replace thermostat(s), bypasses, housing(s), and seals.

To function effectively a thermostat must:

- start to open at a specified temperature.
- be fully open at a set number of degrees above the start to open temperature.
- define a flow area through the thermostat in the fully open position.
- permit zero coolant flow or a defined small quantity of flow when in the fully closed position.

The cooling system thermostat is normally located either in the coolant manifold or in a housing attached to the coolant manifold. Its primary function is to permit a rapid warm-up of the engine: when the engine has attained its normal operating temperature, the thermostat will open and permit coolant circulation. As the thermostat defines the flow area for circulating the coolant, there may be more than one. A heat sensing element actuates a piston that is attached to the seal cylinder. When the engine is cold, coolant is routed to the coolant pump to be recirculated through the engine. When the engine heats to operating temperature, the seal cylinder gates off the passage to the coolant pump and routes the coolant to the radiator. The heat sensing element consists of a hydrocarbon or wax pellet into which the actuating shaft of the thermostat is immersed: as the hydrocarbon or wax medium expands, the actuating shaft is forced outwards in the pellet, opening the thermostat. Thermostats can be full blocking or partial blocking.

Top By-pass Thermostat

The top by-pass thermostat simultaneously controls the flow of coolant to the radiator and the by-pass circuit. During engine warm up, all of the engine coolant is directed to flow through the by-pass circuit. As the temperature rises to operating temperature, the thermostat begins to open and coolant flow is routed to the radiator, increasing incrementally with temperature rise.

Poppet or Choke Thermostats

Poppet type thermostats control the flow of coolant to the radiator only and the by-pass circuit is open continuously. Flow to the radiator is discharged through the top of the thermostat valve.

Side By-pass or Partial Blocking Thermostat

The side by-pass thermostat functions similarly to the poppet type. It has a circular sleeve below the valve which moves with the valve as it opens: this serves to partially block the bypass circuit and direct most of the flow to the radiator.

Vented and Unvented Thermostats

Vented thermostats have a small orifice in the valve itself or a notch in the seat. Usually this must be positioned in a upright position on installation. The function of the vent orifice is to help deareate the coolant by routing air bubbles out of the by-pass circuit. Positive dearation type systems usually require non-vented thermostats.

By-pass Circuit

The term by-pass circuit describes the routing of the coolant before the thermostat opens, that is, through the engine cylinder block and head. The flow of by-pass coolant permits rapid engine warmup to the required operating temperature.

Running without a thermostat is not recommended. It also contravenes the EPA requirements regarding tampering with emission control components. Removing the thermostat invariably results in the engine running too cool. This can cause vaporized water in the crankcase to condense and cause corrosive acids (sulfuric acid is H_2SO_4) and sludge in the crankcase. Additionally, low engine running temperatures will increase the emission of HC. Conversely, on engines that should use top bypass or partial by-pass-type thermostats may overheat when the thermostat is removed as most of the coolant will be routed through the by-pass circuit with little being routed in the radiator. Remember, there is a difference between start to open and fully open temperature values.

Task D9 Flush and refill cooling system; bleed air from system; recover coolant.

Engine coolant is a mixture of water, antifreeze, and supplemental cooling additives (SCA). To cool the engine, coolant must have an unobstructed flow throughout the cooling system. Often when refilling the cooling system, air becomes trapped inside the block. If this condition persists and the engine temperature exceeds 212°F (100°C) water will turn to steam. This steam pressure forces more water out of the block and thereby causes an air lock in the cooling system. To prevent this from happening, manufacturers usually install bleeder valves and provide instructions on their use. Generally, bleeder valves are located near the top of the cooling system and block. After a cooling system has been in operation for several years, its efficiency may be reduced due to formation of scale and sludge. If this is the case, the radiator may need to be removed and flushed. While this can be performed in the vehicle, it generally is not very effective. Drain the coolant system into a container by opening the lower drain valves and upper bleed valves. Coolant should be tested before reusing. Once the radiator or some other component is replaced, refill the system ensuring that the bleeder valves are open. When all the air is out of the system, close the bleeder valves, start the engine, and check the coolant level. Test the antifreeze protection level at operating temperature. In modern engines the antifreeze will prevent corrosion and raise the temperature at which the coolant will boil under pressure. This lengthens the life of the engine and cooling system components.

Task D10 Inspect, repair/replace coolant conditioner/filter, check valves, lines, and fittings.

When freeze protection is required, an antifreeze meeting engine manufacturer specification must be used. An inhibitor system is included in most types of antifreeze and no additional inhibitors are required on initial fill if a minimum antifreeze concentration of 30 percent by volume is used. Solutions of less than 30 percent concentration do not provide sufficient corrosion protection. Concentrations over 67 percent adversely affect freeze protection and heat transfer rates. Most manufacturers suggest a 40 percent antifreeze to 60 percent solution. Ethylene glycol base antifreeze is recommended, such as in all Detroit Diesel engines. Methyl alcohol base antifreeze is not recommended because of its effect on the non-metallic components of the cooling system and because of its low boiling point. Methoxy propanol base antifreeze is not recommended for use by Detroit Diesel due to the presence of fluoroelastomer seals in the cooling system. Before installing ethylene glycol base antifreeze in a unit that has previously operated with methoxy propanol, the entire cooling system should be drained, flushed with clean water, and examined for rust and scale contaminants, etc. If deposits are present, the cooling system must be chemically cleaned with a commercial grade heavy-duty de-scaler. The inhibitors in antifreeze should be replenished at approximately 500-hour intervals or by testing with a non-chromate inhibitor system. Commercially available inhibitor systems may be used to re-inhibit antifreeze solutions.

Supplemental Coolant Additives (SCA)

Supplemental coolant additives are a critical ingredient of the coolant mixture. The actual SCA package recommended by an engine manufacturer will depend on whether wet or dry cylinder liners are used, the materials used in the cooling system components, and the fluid dynamics (high flow/low flow) within the cooling system. The operator may wish to adjust the SCA package to suit a specific operating environment or set of conditions. Abnormally hard water for instance, will require a greater degree of anti-scale protection. Depending on the manufacturer, SCA may be added to a cooling system in a number of ways. The practice of installing the SCA in the system coolant filter is less commonly used today as it resulted in generally higher levels of additives than required and an excess of additives can create problems. Most OEMs suggest testing the SCA levels in the coolant followed by adding SCA to adjust to the required values. Never dump unmeasured quantities of SCA into the cooling system at each PM.

Testing SCA Levels

Generally OEMs recommend that the coolant SCA level be tested at each oil change interval. Additionally, whenever there is a substantial loss of coolant and the system has to be replenished, the SCA level should be tested. Each test provided by an OEM is designed to monitor the SCA package required for their product and cannot generally be used for other OEM products. Also, the test kits usually consist of test strips which must be stored in airtight containers and have expiration dates that should be observed. Coolant test kits permit the technician to test for the appropriate SCA concentration, the pH level, and the total dissolved solids (TDS). The pH level determines the relative acidity or alkalinity of the coolant. Acids may form in engine coolant exposed to combustion gases or in some cases when cooling system metals (ferrous and copper base) degrade. The pH test is a litmus test in which a test strip is first inserted into a sample of the coolant, then removed and the color of the test strip indexed to a color chart provided with the kit. The optimum pH window is defined by each OEM, but normally falls between 7.5 and 11.0 on the pH scale. Higher acidity (readings below 7.5 on the pH scale) readings in tested coolant are indications of corrosion of ferrous and copper metals, coolant exposure to combustion gases and in some cases, coolant degradation. Higher than normal alkalinity readings indicate aluminum corrosion and possibly that a low silicate antifreeze is being used where a high silicate antifreeze is required.

Task D11 Inspect, repair/replace water pump, hoses, and idler pulley.

Most water pumps are belts driven using an idler pulley to maintain belt tension. As belts age they stretch and the idler pulley (on most modern engines) automatically adjusts tension. If the belt is loose check for binding in the idler pulley. Water pumps usually allow a small amount of water to escape around the bearings to cool and lubricate them. Expect a few drops to leak from a factory hole in the bottom of the pump casting. Most water pumps are mounted to standoffs on the engine and require inlet and outlet hoses. More and more manufacturers are mounting them into the front of the block and some require no hose connections. With this approach there is a gasket used between the block and water pump. Any damage to a belt is reason to change it, but most manufacturers say it must be changed if two or more cracks appear in any 2-inch length.

Task D12 Inspect and clean radiator, pressure cap, and tank(s); determine needed service.

The radiator has a pressure control cap with a normally closed valve. The cap, with a number 7 stamped on its top, is designed to permit a pressure of approximately seven pounds (48 kPa) in the system before the valve opens while the cap with a number 9 stamped on its top needs nine psi (62 kPa) in the system before the valve opens. This pressure raises the boiling point of the cooling liquid and permits somewhat higher engine operating temperatures without loss of any coolant from boiling. To prevent the collapse of hosed and other parts which are not internally supported, a second valve in the cap opens under vacuum when the system cools. Use extreme care when removing the coolant pressure control cap. Remove the cap slowly after the engine has cooled. The sudden release of pressure from a heated cooling system can result in loss of coolant and possible personal injury (scalding) from the hot liquid. To ensure against possible damage to the cooling system from either excessive pressure or vacuum, check both valves periodically for proper opening and closing pressures. If the pressure valve does not open between 6.25 psi (43.1 kPa) and 7.5 psi (51.7 kPa) or the vacuum valve does not open at 0.625 psi (4.3 kPa) (differential pressure), replace the pressure control cap. It is recommended by some manufacturers that highway vehicle engines use a minimum 9 psi (62 kPa) pressure control cap. If the pressure valve does not open between 8 psi (55 kPa) and 10 psi (69 kPa) or the vacuum valve does not open at 0.625 psi (4.3 kPa) (differential pressure), replace the pressure control cap.

Task D13 Inspect, repair/replace fan hub, fan, fan clutch, mechanical and electronic fan controls, fan thermostat, and fan shroud.

Fan assemblies must be inspected. Clean the fan and related parts with clean fuel oil and dry them with compressed air. Shielded bearings must not be washed: dirt may be washed in and the cleaning fluid could not be entirely removed from the bearing. Examine the bearings for any indications of corrosion or pitting. Hold the inner race or cone so it does not turn and revolve the outer race or cup slowly by hand. If rough spots are found, replace the bearings. Check the fan blades for cracks. Replace the fan if the blades are badly bent, since straightening may weaken the blades, particularly in the hub area. Remove any rust or rough spots in the grooves of the fan pulley and crankshaft pulley. If the grooves are damaged or severely worn, replace the pulley. Examine and measure the fan hub shaft front and rear journals (industrial engines). Compare against manufacturer's specifications. If the journals are worn excessively, replace the fan shaft. Look for cracks in the adjusting and support bracket castings. When replacement of either the fan shaft or the adjusting bracket is necessary, a new fan shaft and bracket assembly must be used. The current fan shaft rear bearing inner race should be inspected for any measurable wear. Replace the inner race if the outer diameter is less, or they fail to meet manufacturer's specifications. Remember the inner and outer races are only serviced as a rear roller bearing assembly. When installing the rear bearing inner race, press it on the shaft and follow manufacturer's specifications for positioning of the bearing from the end of the shaft. Generally, new fan hub spacers and hubcap replacements of older spacer and cap assemblies differ and should be replaced as a set. The spacers (individually or in combination) also provide a means for setting the different clearances. The spacers have a flange on one side that serves as a pilot for the fan as well as a spacer pilot for the second spacer when two or more spacers are used together. When replacing the former fan hub spacer, be sure and include the new cap. Fan hubs equipped with roller bearings may be modified by adding a grease fitting.

Task D14 Inspect, repair/replace radiator shutter assembly and controls.

Radiator shutters are used to maintain coolant operating temperature in cold climates, and to speed up warmup of the diesel engine. By partially blocking off airflow through the radiator shutters excess cooling of the radiator can be prevented. There are two basic ways to control shutters used on trucks. One method is a thermostatic piston that is mounted on the shutter assembly and reacts to outside temperature and radiator heat. Manufacture of this type is usually limited to inspecting linkage and shutters for freedom of movement, and testing thermostatic piston for proper operation. Another method of opening and closing shutters is to use the thermostat regulated by engine temperature to supply compressed air from the engine compressor. The regulated air pressure is used by a piston on the shutter assembly to counteract spring tension that holds the shutter open. Maintenance of this system will include inspection of all air lines for leaks and blockages, and the shutter piston and thermostat. For proper operation, shutter linkages for binding and wear.

E. Air Induction and Exhaust Systems Diagnosis and Repair (11 Questions)

Task E1 Inspect, service/replace air induction piping, air cleaner, and element; check air restriction.

The air type air filter will incorporate a structural design to cause air to circulate around the filter element. This cyclonic action causes airborne particles (like sand and dust), to be slung against the sides of the canister where they fall and collect in the bottom chamber. This reduces the clogging of the filter element and extends its life. The filter element is impregnated with a resin to trap most of the remaining particles. This type of air filter is the most common and has about a 99.6–99.9 percent efficiency rating.

The oil bath air filter uses a specially designed air filled pan to trap large particles. As intake air is forced through the oil the heavier particles are slowed and fall to the bottom in the oil. A filter chamber is filled with porous material to catch and return airborne oil and particles to the oil bath. This type air filter has about 97 percent efficiency at rated revolutions per minute (rpm). Because the filter relies on the velocity of air flowing through the oil, efficiency is reduced at lower engine rpm.

Excessive restriction of the air inlet will affect the flow of air to the cylinders and result in poor combustion and lack of power. Consequently, the restriction must be kept as low as possible considering the size and capacity of the air cleaner. An obstruction in the air inlet system or dirty or damaged air cleaners will result in a high blower inlet restriction. Check the air inlet restriction with a water manometer connected to a removable fitting in the air inlet ducting (non-turbocharged engine) or the compressor inlet (turbocharged engine). When practicability prevents the insertion of a fitting at this point (non-turbocharged engine), the manometer may be connected to the engine air inlet housing. The restriction at this point should be checked at a specific engine speed. Then the air cleaner and ducting should be removed from the air inlet housing and the engine again operated at the same speed while noting the manometer reading. The difference between the two readings, with and without the air cleaner and ducting, is the actual restriction caused by the air cleaner and ducting. Check the normal air inlet vacuum at various speeds (at no-load) and compare the results with the manufacturer's manual engine operating conditions.

Pulsation can occur in the air intake system when the engine is equipped with an exhaust brake. This is caused by the combination of cylinder pressures and the intake valves opening during the braking period. The pulsation can damage the element and cause dirt to move through the element and enter the engine. The air intake suppressor can help to prevent this problem. The air intake suppressor must be installed into the air intake system between the engine and the air cleaner. Install the suppressor as close to the engine as possible. Use the following to help you to determine the need for the suppressor:

- Air intake suppressors are required for naturally aspirated engines that use a dry element air cleaner.
- Check the design of the air cleaner to find if an intake suppressor is necessary.
- Oil bath air cleaners must have a seal between the oil pump and the air cleaner body.

Task E2 Inspect, repair/replace turbocharger, wastegate, boost sensor, overboost protection valve, and piping system.

A turbocharger is like a high-speed windmill turning a fan. Exhaust gas driven rotor blades in the exhaust side of the turbocharger rotate the compressor blades in the intake side of the compressor. Because the rotor and compressor are connected to a common shaft and built to close tolerances, the assembly can achieve speeds between 60,000 to 100,000 rpm. The resultant increase in air density from this air induction above atmospheric pressure increases compressor speed and output. This causes an overall increase in volumetric efficiency of the engine. At high engine rpm, when efficiency is not a problem, a wastegate is used to bypass a portion of the exhaust gas pressure, thereby preventing over-charging of the cylinder. The output air is less dense because the extremely hot turbocharger superheats the air. To reduce the effects of superheating an aftercooler is installed between the turbocharger and the cylinders. While the commonly used aftercooler is an air-to-air heat exchanger, some engine manufacturers prefer to use an air-to-water heat exchanger, called an intercooler. Regardless of the method used, the effects are the same; there is increased efficiency because cooled air is denser.

The impeller of the exhaust side of the turbo is called the turbine, because it is driven by exhaust gases. The impeller of the intake side is called the compressor, because it increases the available air pressure.

As exhaust gases drive the turbine, the compressor connected to the same shaft is rotated. The rotation of the compressor causes air to be drawn from the outside, through the filter and into the inlet side of the compressor. The process of pressurization causes the air to heat and additional heat is transferred from the turbine through the shaft to the compressor and to the compressed air. This hot pressurized air is sent to the aftercooler where it is cooled and supplied to the intake manifold for delivery to each cylinder in turn. After combustion, the hot expanding gases are supplied through the exhaust manifold to the turbine inlet. The hot gases turn the turbine and exit through the outlet to the exhaust and muffler. Some turbochargers use an electrically or mechanically controlled valve to route a portion of the exhaust gases around the turbine. This valve is referred to as a wastegate or overboosts protection valve, and prevents the turbocharger from overboosting the cylinders at high engine rpm. Modern electronic systems incorporate a boost sensor to monitor the turbocharger and an electronic module to control the wastegate of overboosts valve.

Task E3 Inspect, repair, rebuild/replace engine-driven blowers.

Oil from the engine sump is supplied to the bearings of the rotors for lubrication and cooling. Seals are installed between the seals to prevent vacuum pressure from drawing oil into the blower and blower air pressure from forcing oil from the bearing back into the engine.

Proper air box pressure is required to maintain sufficient air for combustion and scavenging of the burned gases. Low air box pressure is caused by a high air inlet restriction, damaged blower rotors, an air leak from the air box (such as leaking end plate gaskets), or a clogged blower air inlet screen. Lack of power or black or gray exhaust smoke are indications of low air box pressure. High-air box pressure can be caused by partially plugged cylinder liner ports. Check the air box pressure with a manometer connected to an air box drain tube. Check the readings obtained at various speeds with engine operating conditions in the manufacturer's manuals.

Task E4 Inspect, repair/replace intake manifold, gaskets, sensors, and connections.

Intake piping, tubing, and hoses are very durable yet very flexible to allow for movement of the engine. To reduce weight solid piping is usually made of aluminum or ridged plastic. Connections between the solid section of piping and major components are made using rubber sections that are clamped at both ends. When these connections become loose, unfiltered air is allowed to enter the intake system. Airborne abrasive or corrosive elements can then cause damage to critical components. On the other hand, as the rubber or plastic components age or become exposed to extreme heat they will collapse and cause restrictions in the airflow. To provide a visual warning to the operator of problems in the air intake system most manufacturers include an air intake vacuum indicator. This could be as simple as a vacuum tube attached to the intake piping that runs to an instrument panel indicator, or an electronic sensor that is monitored by an electronic module. If the vacuum pressure increases, check for a restriction such as a collapsed pipe or clogged filter. A drop in vacuum indicates a possible leak, clogged vacuum indicator tube, or faulty sensor.

Task E5 Inspect, test, clean, repair/replace intercooler (aftercooler) and charge air cooler assembly.

The aftercooler resembles a radiator except the fin tubes appear to be larger and the inlet and outlet connections are larger. This allows a large volume of air to pass through. As the hot pressurized air passes through the small fin tubes, heat is transferred from the pressurized air to the tubes, then to the fins, and finally to the outside air passing between the fins.

Aftercoolers usually require less removal time and can be inspected visually while still in the vehicle. Wash fins with low-pressure water and straighten as necessary. Inspect for missing or loose fins, dented or cinched tubes, corrosion, or holes in tubes. Leaks can be detected on turbocharger assisted vehicles by running the engine, spraying the aftercooler tubes with soap and water, and watching for bubbles.

An intercooler cools the pressurized air using liquid coolant in the opposite way the radiator works. Heated air is passed through small tubes inside a chamber, usually aluminum, through which coolant is circulated. Heat from the air is transferred to the tubes and on to the coolant, which is cooled by the radiator.

Intercoolers are more difficult to inspect, but typically, the affects will be visible in engine performance. If the engine has been shut down for a while there may be traces of water in the oil or the engine may produce a white smoke when running. The turbocharger may have signs of liquid leakage from its compressor outlet or corrosion on the fins. This would also result in unexplained loss of coolant in the radiator, while the engine is still cold. Often the internal leaking of the intercooler can be misdiagnosed as a cracked head or blown head gasket. One common problem with intercoolers is airlocks. Usually intercoolers are located at the top of the engine. When maintenance is performed on the engine resulting in refilling, the coolant air must be bled from the intercooler coolant chamber. Failure to do so could reduce the effectiveness of the cooler and cause overheating due to steam pressure buildup.

Task E6 Inspect, repair/replace exhaust manifold, piping, mufflers, exhaust back pressure regulator, catalytic converter, and mounting hardware.

Task E7 Inspect, repair/replace preheater/inlet air heater, or glow plug system and controls.

Modern diesel engines use glow plugs that are controlled by the PCM. A glow plug (GP) relay control is used to energize the glow plugs for assisting cold engine startup. Engine oil temperature, battery positive voltage (B+), and barometric pressure (BARO) is used by the PCM to calculate glow plug on time and the length of the duty cycle. On time normally varies between 10 and 120 seconds. With colder oil temperatures and lower barometric pressures, the plugs are on longer. If battery voltage is abnormally high, the duty cycle is shortened to extend plug life. The glow plug relay will only cycle on and off repeatedly when there is a system high voltage condition greater than 16 volts. An opening in the glow plug relay circuit will render the glow plugs inoperative. A short circuit will result in a glow plug always on condition. The glow plug light signal controls the WAIT TO START indicator light located on the instrument panel. When the light goes off, the engine is ready to be started. The light comes on every time a key on reset occurs. On time normally varies between 1 and 10 seconds. The WAIT TO START light on time is independent of the glow plug relay on time because the glow plugs may stay on to improve performance until the engine reaches operating temperature. An open circuit in the glow plug light wiring will result in an inoperative glow plug light. A short circuit will result in a glow plug light always on condition.

Task E8 Inspect, repair/replace ether/starting fluid system and controls.

The principal parts of the starting aid pump are the body, plunger, and the spring loaded ball type inlet and outlet check valves. The pump body is threaded externally at one end for mounting purposes. One end of the plunger is threaded into the operating knob. Two seal rings of oil resistant material are located in grooves at the other end of the plunger. The inlet check valve, which opens on the suction stroke of the plunger and seats under pressure, is located in the side opening of the pump body. The check valves are identified by the number "1/2" stamped on the inlet valve and the number "30" on the outlet valve. An arrow indicating the direction of flow is also stamped on each check valve. Clean the parts with fuel oil and dry them with compressed air. Examine the seal rings for wear or cracks. Replace the seal rings if necessary. The check valves cannot be disassembled; however, they may be cleaned by forcing fuel oil through them with any suitable pump. Inoperative valves must be replaced. The fluid starting aid is designed to inject a highly volatile fluid into the air intake system to assist ignition of the fuel at low ambient temperatures. It essentially consists of a pump and nozzle for injecting the fluid into the air intake and a suitable container for the fluid. The fluid is contained in suitable capsules to facilitate handling. The starting aid consists of a cylindrical capsule container fitted with a screw cap. Inside the container is a sliding plunger like a piercing shaft. From the capsule container, a tube leads from the container to a hand-operated pump and another tube leads from the pump to an atomizing nozzle threaded into a tapped hole in the air inlet housing. The pump may be mounted on the instrument panel or in some other convenient location. The capsule container must be mounted in a vertical position away from such high heat areas as the exhaust manifold, muffler, etc. and should not be located under a hood or in a cab. The atomizing nozzle is screwed into a tapped hole in the air inlet housing. The tank to pump tube should be 3/16 in. OD copper tubing and the pump to nozzle tube 1/8 in. OD.

Task E9 Inspect, repair/replace emergency air induction shut-off system.

The shutdown valve is used to release a door inside the air box of the blower, completely restricting the air intake shortly. The vacuum pressure in the air box side of the rotors increases sharply, and the air pressure to the cylinders drops. Without air for combustion the cylinders fail and the engine stops. When the engine (hence rotors) stops, the vacuum pressure in the air box returns to normal. Oil from the engine sump is supplied to the bearings of the rotors for lubrication and cooling. Seals are installed between the rotors and the bearings to prevent vacuum pressure from drawing oil into the blower and blower air pressure from forcing oil from the bearing back into the engine.

Continual use of the air box shutdown can cause the seals to rupture and oil to leak into the chamber. Excess oil in the chamber can cause the engine to smoke or overspeed.

Continual use of the air box shutdown when the engine is idling or a one time shutdown when the engine is running at full rated load and rpm can cause the seals to rupture.

Depending on where the seal ruptures, the rotor bearings can starve or the engine may start smoking. If it becomes serious enough the blower will draw enough oil through the seals to cause the engine to go into an overspeed condition.

Air shutdown devices are designed to block off air to a diesel engine. Without air supplied to the cylinder there is no compressing heat created to ignite fuel and no air for burning fuel. One type of air shutdown is a valve installed in the air intake system. The air shutdown housing is mounted on the blower. A valve mounted inside of the housing may be closed to shut off the air supply and stop the engine when abnormal operating conditions require an emergency shut down. This valve should only be used in case of an emergency. Constant use of this valve to perform normal shutdown will cause damage to the blowers or cylinder rings. On some turbocharged intercooled engines, air shutdown housings are mounted on the intercoolers, which are mounted on the blowers. Long mounting studs with nuts are used to secure the air shutdown housing and intercoolers to the blowers. To remove the shutdown housing:

- Disconnect and remove the air ducts between the air cleaner and the air shutdown housing.
- Disconnect the control wire from the air shutoff controls.
- Remove the bolts and lock washers which attach the housing to the adapter. Then remove the housing and gasket.
- Remove the bolts and washers which attach the housing adapter to the blower. Then remove the adapter and the blower screen.

To disassemble the air shutdown housing:

- Remove the pin from the end of the air shutoff valve shaft. Then remove the washer or spacer from the shaft and the seal ring from the housing.
- Perform a good inspection of the housing and the valve.
- Clean all of the parts thoroughly, including the blower screen, with fuel oil and dry them with compressed air. Inspect the parts for wear or damage. The face of the air shutoff valve must be perfectly flat to assure a tight seal when it is in the shutoff position.
- Resurface the face area as necessary and replace warped valve plates.
- Some turbocharged engines use a nylon bushing in the valve shaft bore. Examine these bushings for wear and replace them if necessary.

F. Fuel System Diagnosis and Repair (20 Questions)

1. Mechanical Components (10 Questions)

Task F1.1 Inspect, repair/replace fuel tanks, vents, cap(s), mounts, valves, screens, supply, crossover, and return lines and fittings.

In diesel fuel systems, there exists a clear divide between the suction side and the charge side represented by the fuel transfer pump. Most fuel subsystems are of this type and the terms suction circuit and charge circuit are used to describe each. A primary filter is most often located on the suction side of the transfer pump while the secondary filter is located on its charge side. However, there are some fuel systems, notably Cummins PT, where all movement of fuel through the fuel subsystem is under suction. When such a fuel system uses multiple filters the terms primary and secondary tend not to be used.

Fuel Tanks

Fuel is stored on commercial vehicles in fuel tanks. Many diesel fuel management systems are designed to pump much greater quantities of fuel through the system than that required for actually fuelling the engine.

Generally, one half to two thirds of the fuel pumped is returned to the tank. This excess fuel factor varies from a minimal amount to values exceeding 60 percent of pumped fuel; it is used to lubricate and especially to cool high-pressure injection components, especially those exposed directly to the extreme temperatures of engine cylinders. As a cooling medium, the fuel transfers heat from the injection devices to the fuel tank. This provides the fuel tank(s) with a role as heat exchanger.

Fuel tank vents should be routinely inspected for restrictions and should be protected from ice buildup. A plugged fuel tank vent will rapidly shutdown a engine, creating a suction side inlet restriction value the transfer pump will not be capable of overcoming.

To check for the presence of water in fuel tanks, first allow the fuel tanks to settle then insert a probe (a clean aluminum welding rod) lightly coated with water detection paste through the fill neck until it bottoms in the base of the tank: withdraw the rod and examine the water detection paste for a change in color. This test will give some idea of the quantity of water in the tank by indicating the height on the probe the color has changed to. Trace quantities (just the tip of the probe changes color) in fuel tanks are not unusual and will not necessarily present any problems.

Fuel Tank Repair

Generally, OEMs recommend that a defective fuel tank be replaced rather than repaired. However, corroded and punctured fuel tanks are going to be repaired regardless of OEM recommendation so the following suggestions are offered not to endorse fuel tank repair practice but to promote some safety awareness. It is essential to recognize the explosion hazard represented by diesel fuel vapors.

If a metal fuel tank must be welded, all fittings and plugs should be removed and the tank high pressure steamed until the temperature exceeds 950°C throughout or for a minimum of one hour; it should also be remembered that in some jurisdictions, the practice of steaming fuel storage vessels to atmosphere is prohibited. Fuel tanks sectioned with baffles must be directly subjected to the steam within each section. After steaming, the tank should be filled with nitrogen gas before repair welding and/or weld patching. When a gasoline tank is to be repair welded, it should be evacuated with a vacuum pump to a moderate vacuum then filled with nitrogen gas.

When evacuating fuel tanks, check either with the tank OEM for the safe vacuum test value. A high vacuum value may collapse a fuel tank: as a rule, cylindrical tanks will withstand higher vacuum test values than square section tanks.

The practice of welding fuel tanks in position and filled with fuel while arguably safer than welding on a tank filled with fuel vapor, should never be undertaken. A single mistake or metallurgical defect in the tank wall will result in the death of the welder and any other person in the vicinity.

Fuel Tank Sending Units

Most commercial truck fuel subsystems use remote (from the tank) fuel transfer pumps and not assemblies that incorporate the sending unit and a transfer pump. So the fuel sending unit is an integral assembly flange fitted to the tank. It consists of a float and arm connected to a variable resistor whose function it is to control current flow to a cab gauge proportionally with fuel tank level. Fuel sending unit problems can be diagnosed with a DMM (digital multimeter) in resistance mode, by moving the float arm through its arc and observing readings.

Pick-up Tubes

Fuel pick-up tubes are positioned so that they draw on fuel slightly above the base of the tank and thereby avoid picking up water and sediment. Pick-up tubes are quite often welded into the tank; in this case, if they fail the tank must be replaced. Fuel pick-up tubes seldom fail but when they do it is usually by metal fatigue crack at the neck; this results in no fuel being drawn out of the tank by the transfer pump whenever the fuel level is below the crack.

Task F1.2 Inspect, clean, test, repair/replace fuel transfer (lift) pump, pump drives, screens, fuel/water separators/indicators, filters, heaters, and associated mounting hardware.

Fuel Charging/Transfer/Pumps

Fuel charging or transfer pumps are positive displacement pumps driven directly or indirectly by the engine. A positive displacement pump displaces the same volume of fluid per cycle and therefore fuel quantity pumped increases proportionately with rotational speed. Similarly, if a positive displacement pump unloads to a defined flow area, pressure rise can be said to be proportional with rpm increase. On most truck and bus fuel systems, charging/transfer pumps are of the plunger or gear types.

Plunger Type Pumps

Plunger type pumps are often used with port-helix metering injection pumps. They are usually flange mounted to the injection pump cambox and driven by a dedicated cam on the pump camshaft. Single acting and double acting plungers may be used, with the latter type specified in higher output engines requiring more fuel. A single acting plunger pump (a pump with a single reciprocating element such as bicycle pump) has a single pump chamber and an inlet and outlet valve. Fuel is drawn into the pump chamber on the inboard stroke and pressurized on the outboard or cam stroke.

A double acting pump has twin chambers each equipped with its own inlet and outlet valve. On the cam stroke, a two way plunger charges the pump chamber while admitting fuel to the other. On the return stroke, the pump retraction spring reverses the process.

Gear Type Pumps

Gear type pumps are also commonly used as transfer pumps. These are normally driven from an engine necessary drive and are located wherever convenient. Gear pumps will usually have an integral relief valve which will define the peak system charging pressure. Fuel injection systems designed to be charged at pressure values higher than typical, tend to use gear type transfer pumps over cam actuated, plunger pumps. In instances where a gear pump feeds an injection system with no main filter in series, a filter mesh is sometimes incorporated to protect injection pumping apparatus: when gear pumps are used, there is a small chance that gear teeth cuttings can be discharged into the system. It should be noted that a majority of the full authority, electronic management fuel systems use gear type pumps.

Hand Primer Pumps

A hand primer pump may be a permanent fixture to a fuel subsystem located on the fuel transfer pump body or a filter mounting pad. A hand primer pump can be a useful addition to the technician's tool kit and it can be fitted to a fuel subsystem when priming is required. The function of a hand primer pump is to prime the fuel system whenever prime is lost. Typically they consist of a hand actuated plunger and use a single acting pumping principle. On the outward stroke, the plunger exerts suction on the inlet side, drawing in a charge of fuel to the pump chamber; on the downward stroke, the inlet valve closes and fuel is discharged to the outlet. When using a hand primer pump, it is important to purge air downstream from the pump on its charge side. Some fuel subsystems mount a hand primer to the transfer pump housing. Some newer fuel systems have self contained, electric priming pumps whose function is to prime the system after servicing.

Priming a Fuel System

While this is a relatively simple procedure, it is usually advisable to consult the appropriate service manual. Caterpillar suggests pressurizing the fuel tank(s) with regulated air pressure at 35 kPa (5 psi) and cranking the engine.

CAUTION! Never exceed 55 kPa (8 psi) or tank may be damaged.

However, most OEMs prefer that the technician avoid pressurizing air tanks. It should be remembered that diesel fuel contains volatile fractions and the act of pressurizing a fuel tank with air pressure will vaporize some fuel while air exiting an air nozzle creates friction and the potential for ignition. Avoid this practice in extreme hot weather conditions.

Recommended Priming Procedure

When a vehicle runs out of fuel and it is determined that the fuel subsystem requires priming, remove the filters and fill with filtered fuel:

- Locate a bleed point in the system—often on an in-line injection pump system, this will be at the exit of the charging gallery and crack open the coupling.
- Next, if the system is equipped with a hand primer pump, actuate it until air bubbles cease to exit from the cracked open coupling. If the system is not equipped with a hand primer pump, fit one upstream from the secondary filter and actuate until air bubbles cease to exit from the cracked open coupling.
- Retorque coupling. Crank engine for 30 second segments with at least 2-minute intervals between crankings until it starts: this will allow for starter motor cool-down. In most diesel engine systems, the high-pressure circuits will self prime once the subsystem is primed.

Fuel Filters

The function of a fuel filter is to entrap particulate (fine sediment) in the diesel fuel and some current secondary filters will filter to the extent that water in its free state will not pass through the filtering media. A typical fuel subsystem with a suction circuit and a charge circuit will in most cases employ a two filter arrangement, one in each of the suction and charge circuits. Two basic types of filter are used: the currently more common spin-on, disposable cartridge type and the canister and disposable element type. Spin-on filters are obviously easier to service and are the filter design of choice by most manufacturers.

Primary Filters

Primary filters represent the first filtration stage in a typical two stage filtering, fuel subsystem. Primary filters are therefore usually under suction, plumbed in series between the fuel tank and the fuel transfer pump. They are designed to entrap particulate sized larger than 20–30µ (1µ = one millionth of a meter) depending on the fuel system, and achieve this using media ranging from cotton threaded fibers, synthetic fiber threads and resin impregnated paper.

Secondary Filters

Secondary filters represent the second filtration stage in two stage filtering. In a typical fuel subsystem, the secondary filter is charged by the transfer pump and this enables use of more restrictive filtering media. The secondary filter would therefore normally be located in series between the transfer or charging pump (the pump responsible for pulling fuel from the fuel tank and charging the fuel injection components) and the fuel injection apparatus. Current secondary filters may entrap particulate sized as small as 1µ (one millionth of a meter) but filtering efficiencies of 2–4µ are more common.

Water in its free or emulsified states will not be pumped through many of the current generation of fuel filters. This results in the filter plugging on water and shutting down the engine by starving it for fuel. Secondary filters use a variety of media including chemically treated pleated papers and cotton filters.

In a fuel subsystem that is entirely under suction such as Cummins PT., the terms primary and secondary are not used to describe multiple filters when fitted to the circuit. As every filtering device used in the fuel subsystem is under suction, the inlet restriction specification is critical and if it exceeded the maximum, it would result in a loss of power caused by fuel starvation.

Servicing Filters

Most fuel filters are routinely changed on preventative maintenance schedules that are governed by highway miles, engine hours, or calendar months. They are seldom tested to determine serviceability. When filters are tested, it is usually to determine if they are restricted (plugged) to the extent they are reducing engine power by causing fuel starvation.

Primary filters or a filter under suction should be tested for inlet restriction with a mercury (Hg) filled manometer. A manometer is a clear tubular column formed in a U-shape around a calibration scale marked in inches: the column is then filled with either mercury or water (usually colored with dye for ease of reading) to a zero point on the calibration scale. When the manometer is connected to a fluid circuit, it will produce a reading according to the pull (vacuum circuit) or pressure acting on the fluid in the column. Actual inlet restriction values vary considerably with the fuel system and specifications should always be referenced. The Hg manometer should be plumbed between the filter mounting pad and the transfer pump. Transfer pumps are usually positive displacement (unload a constant slug volume of fluid per cycle), so they pump more fuel proportionally with rpm increase; this means valid test results can be obtained without loading the engine. When testing circuit restriction on a fuel subsystem that is entirely under suction such as the Cummins PT, the specifications are likely to be fairly exacting. Circuit restriction specification readings on Cummins PT should be 4–6 in. Hg with a new filter but 7 in. Hg is the maximum specification.

When this value is exceeded, it will result in fuel starvation to the PT pump. The technician should note that most filters function with optimum efficiency just before they completely plug, in other words, at the end of their service life.

The transfer pump usually charges secondary filters. Testing charging pressure (the pressure downstream from the charging/transfer pump) is normally performed with an accurate, fluid filled pressure gauge plumbed in series between the transfer pump and injection pump apparatus; it is not generally used as a method of determining the serviceability of a secondary filter. Secondary filters tend to be changed by preventative maintenance schedule rather than by testing or when they plug on water and shutdown an engine.

Procedure for Servicing Spin-on Filter Cartridges

Much dirt is placed in diesel fuel systems by technicians using improper service techniques. Most diesel service technicians realize that sets of replacement filters should be primed, that is, filled with fuel before installation, but few concern themselves about the source of the fuel. Filters should be primed with filtered fuel. Shops performing regular engine service should have a reservoir of clean fuel; any process that requires the technician to remove fuel from vehicle tanks will probably result in it becoming contaminated at least to some extent however much care is exercised. The container used to transport the fuel from the tank to the filter should be cleaned immediately before it is filled with fuel. Paint filters (the paper cone shaped type) can be used to filter fuel. The inlet and outlet sections of the filter cartridge should be identified. The filter being primed should be filled only through the inlet ports usually located in the outer annulus (ring) of the cartridge and never directly into the outlet port, usually located at the center.

Some manufacturers prefer that only the primary filters be primed prior to installation during servicing. After the primary filter(s) has been primed and installed, the secondary filter should be installed dry and primed with a hand primer pump.

Water Separators

Most current diesel engine powered highway vehicles have a fuel subsystem with fairly sophisticated water removal devices. Water appears in diesel fuel in three forms, free state, emulsified, and semi-absorbed. Water in its free state will appear in large globules and because of its greater weight than diesel fuel, will readily collect in puddles at the bottom of fuel tanks or storage containers.

Water emulsified in fuel appears in small droplets; because these droplets are minutely sized, they may be suspended for sometime in the fuel before gravity takes them to the bottom of the fuel tank. Semi-absorbed water is usually water in solution with alcohol, a direct result of the methyl hydrate (type of alcohol added to fuel tanks as deicer or in fuel conditioner) added to fuel tanks to prevent winter freeze-up. Water that is semi-absorbed in diesel fuel is in its most dangerous form because it may emulsify (the fine dispersion of one liquid into another) in the fuel injection system where it can seriously damage components.

Generally, water damages fuel systems for three reasons. Water possesses lower lubricity than diesel fuel, has a tendency to promote corrosion, and its different physical properties effect the pumping dynamics. Diesel fuels are compressible at approximately 0.5 percent per 1000 psi.; water is less compressible at approximately 0.35 percent per 1000 psi. Fuel injection pumping apparatus is engineered to pump diesel fuel and if water with its lower lubricity and compressibility is pumped through the system, the resultant pressure rise can cause structural failures, especially at the sac/nozzle area of the fuel injectors.

Often a water separator will combine a primary filter and water separating mechanism into a single canister. These use a variety of means to separate and remove water in free and emulsified states; they will not remove water from fuel in its semi-absorbed state.

Water separators use combinations of several principles to separate and remove water from fuel. The first is gravity. Water in its free state or emulsified water that has been coalesced (where small droplets come together to combine into larger droplets) into large droplets will because of its heavier weight, be pulled by gravity to the bottom of a reservoir or sump. Some water separators use a centrifuge to help separate both larger globules of water and emulsified water from fuel; the centrifuge subjects fuel passing through it to centrifugal force, throwing the heavier water to the sump walls where gravity can pull it into the sump drain. A centrifuge will act to separate particulate from the fuel in the same manner. Fuel directed through a fine resin coated, pleated paper medium will pass through the medium with greater ease than water. Water entrapped by the filtering medium can collect and coalesce in large enough droplets to permit gravity to pull it down into the sump drain.

To find a water separator restriction, a mercury (Hg) manometer is the appropriate test tool. In cases where the entire fuel subsystem is under suction, the consequences of exceeding the restriction specification are generally more severe, the result being fuel starvation to the engine.

Servicing water separator units is a simple process but one that should be undertaken with a certain amount of care as it is easy to contaminate the fuel in the separator canister either by priming it with unfiltered fuel or by permitting dirt to enter when the canister lid is removed. Most water separators have a clear sump through which it is easy to observe the presence of water. All water separators are equipped with a drain valve; the purpose of this valve is to siphon water from the sump. Water should be routinely removed from the drain valve. The filter elements used in combination water separator/primary filter units should be replaced in most instances with the other engine and fuel filters at each full service. However some manufacturers claim their filter elements have an in-service life that may exceed the oil change interval by two times. Whenever a water separator is fully drained, it should be primed before attempting to start the engine.

Fuel Heaters

In recent years, it is more common to find trucks equipped with fuel heaters. In fuel systems where fuel is flowed through the injection system circuitry at a rate much higher than that required for fueling the engine, constant filtering of fuel removes some of the wax and therefore some of its lubricity even when the appropriate seasonal pour point depressants are present.

An electric heating element uses battery current to heat fuel in the subsystem. This type offers a number of advantages, most notable of which is that the heater can be energized before startup so that cranking fuel is warmed up. Electric element fuel heaters may be thermostatically managed so that fuel is only heated as much as required and not to a point which compromises some of its lubricating properties.

Fuel heaters exist that use both electric heating elements and coolant medium heat exchangers and furthermore, manage the fuel temperature. Optimally, fuel temperature should be managed not to exceed 900°F (320°C). Once fuel exceeds this temperature its lubricating properties start to diminish and the result is reduced service life of fuel injection components

Task F1.3 Check fuel system for air; determine needed repairs.

You check for air in the fuel system by installing a clear plastic hose in the return line and look for consistent large bubbles. To troubleshoot the source of air admission to the fuel subsystem, a diagnostic sight glass can be used. It consists of a clear section of tubing with hydraulic hose couplers at either end and it is fitted in series with the fuel flow. However, the process of uncoupling the fuel hoses will always admit some air into the fuel subsystem, so the engine should be run for a while before reading the sight glass.

Task F1.4 Prime and bleed fuel system; check, repair/replace primer pump.

A fuel system must be sealed to operate properly. Any introduction of air can cause air locks, which prevent fuel from reaching the injector. To clean air from the line after performing maintenance, manufacturers install primer pumps to pressurize the system. Once pressurized, opening the end of the lines will force trapped air out. Some injectors incorporate a return line that allows excess fuel to return to the tank. This type of system is self-priming, but some manufacturers strongly recommend against priming this way. The excess fuel is used to cool and lubricate fuel system components. Prolonged operation of the engine without fuel will cause fuel system failures or excess wear. Most manufacturers will install a hand primer to make priming systems easier. One type used is a single acting plunger type supply pump. This not only has a hand primer pump that works like a tire pump, but also has a filter/strainer and sediment bowl.

To prevent the engine from starting while performing these checks, disconnect the ignition and tachometer harness from the distributor on gasoline engine models. On diesel engine models, disconnect the wire lead at the fuel shutoff solenoid. Although the starter motor cannot be checked against specifications on the vehicle, a check can be made for excessive resistance in the starting circuit. First, make sure the battery is fully charged. If the battery was recently charged, remove the surface charge. Then, perform a load test. The vehicle should be at room temperature before performing these checks. Colder weather will cause higher resistance in the circuit. Do not operate the starting motor continuously for more than 30 seconds without pausing to allow it to cool at least two minutes. Overheating from excessive cranking will seriously damage the cranking motor. Connect a voltmeter across the battery. It should show at least 12 volts. Then, crank the engine and check the voltage. If there is more than a 2-volt drop, there is excessive resistance in the circuit. Check the ground circuit for loose, corroded or dirty connections, or frayed cables. Check the starter ground by connecting a voltmeter from the negative terminal of the battery to the ground stud on the starter motor field frame. Crank the engine. The voltmeter should read about 0.2 volt. If the reading is 0.5 volt or higher, there is excessive resistance in the ground circuit. If the circuit wires and connections and the ground circuit test as good, but there is still excessive resistance, remove the starter motor for repair.

Task F1.5 Inspect, adjust, repair/replace throttle linkage, and controls.

Throttle linkage adjustments have become extremely critical on today's engine. The travel of the throttle controls the engine over its complete operating range. For example, on the Cummins PT fuel system with a PTG type pump, just setting the end stops is a touchy job. On governors using air fuel control (AFC), perform throttle linkage adjustments according to manufacturer's instructions. Then make sure the throttle shaft is all the way back against the rear throttle stop screw before adjusting the lever position. Do not move the rear stop screw as this will change the throttle linkage adjustment previously made. Check the travel from idle to the full fuel position; this must meet manufacturer's calibration specifications. If it is incorrect, adjustment of the front stop screw will correct the problem. Then, check and adjust the idle speed according to manufacturer's procedures. Before making any critical adjustments, ensure proper fuel pressure and flow is available. Check fuel filters for obstructions, sticking injector and throttle linkages, low fuel pressure from worn pumps, and chinked throttle. Improper setting of the throttle can be detected when checking the engine response to load changes. When the throttle is not responding to changes in demand, leading to a too rich mixture of fuel, this will cause a sudden loss of power or lugging.

Task F1.6 Perform on-engine inspections, tests, adjustments, and time, or replace and time distributor-type injection pump.

The distributor-type pump and injection nozzle system is distinctive in appearance because the pump looks like a distributor. In fact, it operates much like an electrical distributor in that it distributes fuel to each cylinder when it is ready to fire. An internal valve is rotated to line up with a high-pressure line dedicated to only one of the injectors. When the valve and line are aligned fuel flows to that injector and it fires. Further rotation aligns the valve with the next injector's line and the next injector fires. This process is repeated until all injectors have fired and it starts again. Injector timing is accomplished by routing the injector lines to each injector in firing order, much like spark plug wires are attached in firing order. Fuel is metered and pressurized in the injector pump. Actually, the injector only atomizes the fuel into the hot cylinder for combustion, as the spark plug only sparks to ignite the fuel for combustion. To cool the injector tips, a portion of the fuel is returned to the pump by a nozzle leak-off line.

To remove an internal shutdown and/or cold advance solenoid you remove the pump cover, then the terminal contact nuts and wires from the solenoid.

Remove the solenoid from the cover, clean the gasket material from the cover, and pump. Note the positioning of insulating washers. To install an internal shutdown and/or cold advance solenoid:

- Install insulating washers to the correct terminals and insert the solenoid into the cover.
- Connect wires and terminal contact nuts.
- Replace the cover gasket.
- Install the cover on the fuel pump.
- Test the solenoid for proper operations.

Task F1.7 Perform on-engine inspections, tests, adjustments, and time, or replace and time in-line type injection pump and drives.

The in-line injector pump performs the function of metering, pressurizing, timing, and delivery of fuel for injection. The pump is normally gear driven from the engine camshaft. This, in turn, causes an internal camshaft to operate plungers in much the same manner as the engine camshaft operates valves. The fuel pressure developed by the plunger unseats a valve in the top of the barrel, and fuel is delivered to the injector.

The typical port-helix metering pump used to fuel a truck is in-line configured and flange mounted to an engine accessory drive to be driven through one complete rotation (3,600) per complete engine cycle (7,200). The internal pump components are housed in a frame constructed of cast aluminum, cast iron, or forged steel. The engine crankshaft drives the injection pump by means of timed reduction gearing. The gear driven pump drive plate is connected to the injection pump camshaft, (the shaft fitted with eccentrics designed to actuate the pump elements), so rotating the pump drive plate rotates the pump camshaft. The camshaft is supported by main bearings and rotates within the injection pump cambox. The cambox is the lower portion of the injection pump, which houses the camshaft, tappets, and the integral oil sump. Timing of the pump camshaft to the engine camshaft is critical. Rough timing of high-pressure fuel lines determines injector-firing order.

Timing Injection Pumps to an Engine

Port-helix metering injection pumps are timed to the engine they manage by phasing port closure on #1 cylinder; check the specification (for a variety of reasons, the pump may be required to be timed to #6 cylinder) to a specific engine position. All injection pumps must be accurately timed to the engine they will fuel. This usually means the phasing of pump port closure to a specific number of degrees BTDC on the cylinder to be timed to a specification that seldom can be outside of 10 crank angle and may have to be within 1/40 crank angle. Methods used to time the injection pump to the engine vary by OEM. Actual spill timing of the pump to the engine is a procedure that has become effectively obsolete; however, it is important that the diesel technician understand this procedure which can be used in the absence of OEM tooling.

SERVICE TIP: Most injection pumps on North American engines are timed to #1 engine cylinder but not all, so watch for those that are not. #6 cylinder (on an in-line 6) is next most common but assume NOTHING. Always check the specifications in the shop manual.

Spill Timing Procedure

Prior to beginning the spill timing procedure, check the engine (and OEM chassis) manual for positioning of the fuel control lever, stop fuel lever, brake valve, and gear shift lever position.

Check injection pump data plate for port closure value. Manually bar engine in direction of rotation to position piston within #1 cylinder on its compression stroke. Locate engine calibration scale, usually found on front pulley, harmonic balancer, or on the flywheel. Position engine roughly 200 before the port closure spec.

1. Remove high-pressure pipe from the delivery valve on the injection pump #1 cylinder. Unscrew the delivery valve body and remove delivery valve core and spring. Replace the delivery valve body and install a spill tube. A discarded high-pressure pipe neatly cut and shaped to a goose neck will suffice. When the hand primer pump is actuated, the charging gallery will be pressurized. The amount of pressure created by the hand primer pump will be insufficient to open the delivery valves, so it will exit through the spill tube fitted to the #1 cylinder. The fuel should exit in a steady stream and it should be captured in a container held under the spill tube. Next, slowly and smoothly bar the engine in its direction of rotation observing the stream of fuel exiting the spill tube.

2. When the plunger leading edge rises to trap off the spill port, the steady stream of fuel exiting the spill tube will break up first into droplets and then cease as the plunger passes the spill port.

3. This step locates the pump precisely at port closure—this means that the flow area at the spill port should exist but be minimal; 2–6 drops per 10 seconds should be firmly in the specification window. Ensure that the pump has not been barred past port closure cutting the fuel off altogether as it is impossible to determine how much beyond port closure the plunger has traveled.

4. Next, check the engine calibration scale. The specification is typically required to be within 10 crank angle of the PC spec. If this is so, the pump can be assumed to be correctly timed to the engine. If it is not, proceed as follows.

5. Remove the accessory drive cover plate. Loosen the fasteners that couple the pump drive gear to the pump drive plate. Bar the engine to the correct PC spec position. By uncoupling the pump drive plate from the pump drive gear, it is hoped that the pump will remain stationary at PC on #1 cylinder while the engine is barred independently. When the engine is in the correct position, torque the fasteners that couple the pump drive gear to the pump drive plate. Back the engine up roughly 200 before the PC spec and then repeat steps (1) through (3).

Every diesel technician should be acquainted with the above procedure even though he/she may seldom practice it. Spill timing can also be performed using compressed air as the test medium: the set-up procedure is the same as above, but regulated compressed air is ported into the charging gallery. Instead of a spill tube, a flexible hose is connected to the pump element used for the spill timing procedure and immersed in a water filled glass jar: when air is supplied to the charging gallery, it will pass through the barrel spill ports and exit through the flexible hose producing bubbles in the glass jar. As the pump is rotated toward the port closure location and the plunger leading edge starts to trap off the spill port, the stream of bubbles will turn into smaller bubbles produced less frequently. Using air to spill time injection pumps is probably the least recommended method. The air supply should always be equipped with an air dryer/filter assembly: and even when it is so fitted, the danger of moisture and other air suspended contamination does not warrant the risk of using this method. OEMs prefer alternative methods of static timing of port-helix metering injection to be used that do not require the removal and disassembly of the delivery valve. Some of the alternatives:

Timing pin. This is probably the simplest method and least likely to present problems. The engine is located to a specific position by inserting a timing bolt usually in the cam gear but sometimes in the flywheel. Similarly, the injection pump is pinned by a timing tool to a specific location. The injection pump is always removed and installed with the timing tools in position. It goes without saying that the timing tools must be removed before attempting to start the engine.

High-pressure pump. This involves connecting a high-pressure timing pump into the circuit. This portable electric pump charges the charging gallery at a pressure in excess of the 20 atmospheres required to crack the delivery valves which will cause them all to open; consequently, the injector leak-off system and pump gallery return must be plugged off. Next, the high pressure pipe on #1 cylinder is removed at the pump, and a spill pipe discharging into the portable pump sump is fitted. The procedure then replicates that used to spill time the injection pump—the engine must be correctly located well before the PC spec (to eliminate engine gear backlash variables) and then barred to spill cut-off.

If the settings are out of specification, they are rectified by altering the coupling location of the pump drive gear to the pump drive plate.

Task F1.8 Perform on-engine inspections, tests, and adjustments, or replace pressure timed injection pump, drives, and injectors.

The original pressure timed (PT) pump used a regulator to control the maximum fuel manifold pressure. Later models incorporated a governor control and referred to it as PTG. The most recent improvement (PTG AFC) uses a flow-no-flow by-pass valve. This improved version (AFC) uses manifold vacuum air to control fuel flow during periods of low intake manifold pressure. Regardless of the type of PT pump used, they will be comprised of a gear pump to transfer fuel from the tank and a pulsation damper to absorb pulsation is caused by operation of the gear pump. Additionally, the PT pump uses a MIN-MAX governor to control engine speed between low and high idle or to allow the operator to adjust engine speed.

Task F1.9 Perform on-engine inspections, tests, and adjustments, or replace unit injectors.

In a mechanical unit injector fueled engine, each engine cylinder has its own unit injector, essentially a rocker actuated, pumping and metering element and hydraulic injector nozzle combined in one unit. Each mechanical unit injector (MUI) has fuel delivered to it at charging pressure, variable between 30–70 psi (200 kPa–470 kPa) dependent on engine speed.

Charging pressure is generated by a positive displacement gear pump, driven by the roots blower drive shaft, and responsible for all movement of fuel through the fuel subsystem. Engine output is controlled by a mechanical governor (in most truck and bus applications) that regulates fuel quantity by controlling the unit injector fuel racks.

Jumper pipes connect the fuel manifolds in the head with the unit injectors; one charges the unit injector, the other returns fuel to the return manifold. Fuel is cycled through the system whenever the engine is running. In other words, all of the fuel delivered to the MUI is not used for fueling the engine. Fuel cycled through the MUI circuitry is used for cooling and lubricating the internal components of the assembly.

Mechanical Unit Injectors (MUI)

A mechanical unit injector combines a complete pumping, metering element, and a hydraulic injector nozzle in a single, cam actuated unit. There are two types differentiated by nozzle valve design, crown valve and needle valve. However, the crown valve type with NOP values of 800 psi can be considered obsolete and will not be discussed here. The needle valve type was introduced in the early 1970s and is almost universal in DDC mechanical engines in current use.

Operation

For most of the cycle when the injector train is riding on the inner base circle of its actuating cam, the MUI plunger will be retracted by the injector spring, which also serves to load the injector actuating train, extending from the rocker to the cam follower. Fuel will flow into the MUI from the supply jumper pipe, pass through the lower bushing port charging the pump chamber, pass up through the plunger center and cross drillings, charging the recessed metering helices and exit at the upper bushing port. From the upper bushing port, the fuel flows through ducts to exit the MUI by means of return jumper pipes.

Actual plunger stroke is defined by cam profile geometry and therefore will not vary. At each rotation of the camshaft, each MUI plunger is driven through a single stroke that begins when the cam ramps off its base circle toward peak lift. Cam peak lift represents the maximum point of downwards travel of the plunger into the bushing pump chamber. After peak lift, the cam ramps back toward cam base circle and the injector spring lifts the plunger back into its retracted position.

Fuel Injector Timing

The injector timing dimension specification can be found stamped on valve/rocker cover plate or in Section 14 of the DDC service manual. A timing dimension pin with the correct dimension stamped on its flange should be selected. The engine may be rotated in the same manner as when performing the valve adjustment and there is no reason why both adjustments cannot be set simultaneously, observing correct engine position and sequencing. Set the timing dimension when the valves in the cylinder are fully depressed. Insert the dowel in the timing gauge into the injector timing hole located in the upper flange of the MUI. Smear a drop of engine oil on the injector follower flange and then adjust the injector pushrod so that when the timing tool is rotated it should gently wipe the oil film from the follower. The injector actuating train consists of a solid pushrod that connects to the rocker by means of a pivot pin and clevis: the pushrod connects to the clevis by means of screw threads and a locknut. An adjustment is therefore performed by backing off the locknut and either reducing or lengthening the pushrod total length to set the specified plunger travel. All the injectors are set in the same way, that is, setting actual plunger travel by adjusting the MUI actuating train dimension.

Governor Gap Adjustment

This step is critical to engine performance and if not properly set, the result will be a lack of power and fueling flat spots. The procedure varies somewhat according to engine series and governor type. On 71 series in-line SWLS the gap is set with the engine stopped. With DWLS governed in-lines, the service manual outlines setting the governor gap with the engine running but later service bulletins recommend an engine stationary method, using a wedge tool to spread the low speed weights. Ensure that the correct procedure is used by consulting the DDC service literature.

Governor gap adjustment procedure—engine not running:

1. Remove the governor cover and back out the buffer screw and the starting aid screw, if equipped.

2. Next, bar the engine to position the governor weights in a horizontal position, then insert a governor weight fork (the common practice of using a screwdriver to spread the weights is not recommended in DDC service bulletins) down between the inner lobe of either one of the low speed governor weights (the larger weights) and the weight riser shaft. Use the governor weight fork to the force weights outwards to their fully extended

position against the stop on the weight carrier shaft. Hold the weight fork firmly in this position to maintain it.

3. The preceding step will have forced the low speed spring cap against the high speed plunger. Check DDC specifications to select the correct thickness gauge and insert it at the governor gap; there should be a slight drag. Adjust if required.

4. Reinstall governor cover. If the governor gap setting was checked cold, the adjustment should once again be checked with the engine at operating temperature.

Governor gap adjustment procedure—engine running:

DDC have technical service bulletins on the subject of governor gap adjustments with the engine running. If the service manual procedure is used, the final governor gap adjustment should be checked with the engine at operating temperature. The procedure differs for tailored torque (TT) engines. As always, specifications should be consulted but typically the procedure would be as follows:

1. Non-fuel squeezer engines: the engine should be run at between 800–1000 rpm and the governor gap should be .0015 in. When using shimstock cut with mechanics shears or scissors, ensure that the edge has not been distorted to a much greater than specified dimension.

2. Fuel squeezer (TT) engines: the engine should be run at between 1,100–1,300 rpm and the governor gap should be .002 in.–.004 in.

Set Injector Racks

This is a critical adjustment that must be precisely executed to balance the fueling to each engine cylinder. Erratic setting of the MUI racks during tune-up is a main cause of hunting, a rhythmic fluctuation in engine rpm. The exact linear position of each MUI rack is determined by its control lever adjustment. The governor's location on the engine will determine which injector rack should be set first; this is always that closest to the governor. The technician who is more familiar with Caterpillar 3114/3116/3126 overhead adjustment procedure should be especially alert to the fact that the rack position for full-fuel is reversed on DDC engines, that is, moved to the fully inboard position of travel for full fuel.

1. Disconnect all linkages connected to the governor speed control lever.

2. Back out the idle speed adjusting screw until there is no tension on the low speed spring (LS governors). When backed out approximately 1/2 in. (12–13 threads), the low speed spring will have little or no tension permitting the closing of the low speed gap without risk of bending the fuel rods or damaging the yield link spring mechanism (used with throttle delay mechanisms). Additionally, failure to back out the idle speed adjusting screw will result in inaccurate fuel rack settings.

3. Loosen the adjusting screw (inner and outer adjusting screws on older, two screw rack lever adjusters) of each MUI rack control lever on the control tube. Ensure that the foot of the rack lever is not binding into the MUI rack clevis. More recent model engines use a single adjusting screw and lock-nut with a torsion spring instead of the two screw rack adjusters, to prevent jamming a whole bank into the full fuel position in the event of a single MUI rack seizure. In other words, each rack control lever is capable of returning to its no fuel position independently. On V configuration engines, free the control levers on both banks.

4. On V configuration engines, remove the clevis pin from the right bank fuel rod connection to the control tube, leaving the left bank control tube connected to its fuel rod.

5. On engines with VS mechanical governors, the speed control lever should be pinned into the full-fuel position while setting the injector racks and the stop lever should be held in run position. On LS mechanical governors, load the speed control lever into its full fuel position with light finger pressure.

6. Next, with both LS and VS governed engines, observing the previous step, turn CW the inner adjusting screw (2 screw type) or adjusting screw (one screw type) on the injector closest to the governor until the rack control clevis can be observed to kick-up: at this point, a slight increase in effort required to turn the screwdriver should be

detected. On the single adjustment screw type turn screw CW on further 1/8 turn, then lock in position with the locknut. On the older 2 screw adjustment rack levers, turn the inner screw a further 1/8 turn CW, then screw in the outer screw until it bottoms. Next, alternately tighten each of the screws by turning 1/8 turn CW, finally torquing to specification (24–36 lb/in).

7. Next, on LS governor engines, move the speed control lever through its travel arc, or likewise, the stop lever on VS engines through its arc: if any drag is detected, the rack should be readjusted.

8. On all engines, the first MUI rack to be adjusted is the one closest to the governor; this becomes the master rack once adjusted according to the procedure outlined in the previous step. The remaining rack control levers are adjusted as follows.

9. In-line engines: remove the fuel rod clevis pin from the control tube actuating arm, then hold the injector control racks in the full fuel position using the control tube. Adjust each MUI rack lever in sequence moving away from the master rack down the cylinder bank. Check to ensure that each rack clevis has equal spring-back when tested with the screwdriver. Never readjust the master rack once this is properly set.

Task F1.10 Inspect, test, adjust, and repair/replace fuel injectors.

Unlike the gasoline engine which utilizes an electrical spark to ignite a mixture of fuel and air, the diesel engine must rely solely upon the intense heat of compression, high turbulence in the combustion chamber, and extremely fine atomization (breaking up) of the fuel to obtain ignition. The type of nozzle to be selected depends to a great degree on the type of combustion chamber used in the engine. Generally, indirect (precombustion) combustion chamber engines use pintle nozzles and direct injection engines use orifice or hole type nozzles. The selection of injection nozzles is critical as injection nozzles greatly affect the proper burning of the fuel. Engine manufacturers have placed added emphasis on nozzle design as reduced exhaust emissions (air pollution) become increasingly important to our society. Injection nozzles are simple hydraulic valves operated by fuel pressure. Fuel flow generated by the injection pump enters the nozzle holder at the fuel inlet and proceeds down the feed channel and into the nozzle sac chamber. When the pressure of the fuel against the annular area of the needle valve exceeds the preset pressure of the pressure spring the needle valve is raised from its seat. Then a metered amount of fuel is injected through the orifices on a hole type nozzle or by the pintle on a pintle type nozzle and into the combustion chamber.

When the truck technician refers to an injector, the nozzle holder assembly is generally being described. However, the term is also used perhaps somewhat incorrectly, to describe components such as MUI (mechanical unit injectors), EUI (electronic unit injectors), HEUI (hydraulically actuated electronic unit injectors) and other components which, while they may incorporate a hydraulic nozzle, also perform the metering and timing functions of the diesel fuel system. For the sake of clarity, this text will refer to integral injector assemblies whose function it is to define nozzle opening pressure (NOP) and atomize fuel as hydraulic injectors. It should be noted that one OEM refers to these devices as mechanical injectors (Cummins).

Hydraulic injectors are normally classified by nozzle design. They are simple hydraulic switch mechanisms whose functions are to atomize and inject fuel into the engine cylinders. There are three basic types of injector nozzle, two that are effectively obsolete in medium and heavy-duty truck and bus engine applications. Both will however be briefly described in this section. Poppet and pintle nozzles are best suited to indirect injected engine applications and therefore are not used in any current North American medium and heavy-duty engines. Multi-orifice nozzles are used in most DI (direct injected) diesel engines using in-line, port-helix metering, injection pumps, and as an integral sub-component of mechanically and electronically controlled, unit injection systems.

All current, high-speed diesel engines found in highway applications use hydraulic nozzles with the exception of the open nozzle designs used in Cummins hydromechanical and electronic common rail systems. Open nozzle injectors are covered in later sections of this textbook. There are almost no similarities in operating principle between the closed hydraulic nozzles addressed in this chapter and open nozzle injectors.

Nozzle Opening Pressure (NOP)

An injector nozzle is a hydraulic switch. One of its primary functions is to define the pressure required to trigger its opening. The term popping pressure is also used to describe the opening pressure of a nozzle. The actual NOP value is defined by the mechanical spring tension of the injector spring. This spring tension loads the nozzle valve onto its seat and therefore determines the hydraulic pressure required to unseat the valve. Most injectors incorporate a means of adjusting the injector spring tension so that the NOP value can be set to specification. The spring tension adjustment mechanism is either shims or an adjusting screw and locknut. The NOP value is always one of the first performance specifications to be evaluated when testing injector nozzles on a bench test fixture (pop tester).

Poppet Nozzles

Caterpillar used poppet nozzles in the past on IDI (indirect injection) engines using a pre-combustion chamber. They are perhaps the most simple of the hydraulic injector nozzles and as such were the cheapest to manufacture. Poppet nozzles are not easily reconditioned; the technician is normally required to bench test the nozzle for the correct NOP value and observation of spray pattern and either reject or accept it for continued service.

Poppet nozzles use an outward or forward opening valve principle. This means that all the fuel pumped to the injector ultimately ends up in the engine cylinder eliminating the necessity for the leak off lines (return circuit) used with most other nozzle types. Nozzle opening pressure parameters generally range between 35–125 atms (500–1800 psi) for instance, the Caterpillar poppet nozzles had NOP of around 55 atms (800 psi).

Poppet nozzles operate as hydraulic switches to define the NOP and atomize the fuel. The nozzle spring loads the poppet valve onto its seat. Hydraulic line pressure (from the injection pump) acts on the sectional area of the seated upper portion of the poppet valve and when the injection pump delivers pressure rise sufficient to unseat the poppet valve, fuel passes around the poppet to exit from the nozzle's single orifice.

Fuel is delivered to the pintle injector from the fuel injection pump by means of a high-pressure pipe. The nozzle valve is held in the closed position mechanically by spring pressure either acting directly on the pintle valve or transmitted by means of a spindle (shaft that transmits spring force to the nozzle valve). Fuel is ducted through the injector assembly to the pressure chamber within the nozzle: this permits the line pressure to act on the sectional area of the pintle valve exposed to the pressure chamber. Whenever the hydraulic pressure generated by the injection pump acting on the sectional area of the pintle valve exposed to the pressure chamber is sufficient to overcome the spring pressure loading the valve on its seat it retracts, permitting fuel to flow past the seat and exit through the injector orifice. Fuel will continue to flow through the nozzle orifice for as long as hydraulic pressure exceeds the mechanical force represented by the injector spring. The fuel injection pump determines the length of a fuelling pulse and when line pressure collapses, the spring reseats the nozzle valve, ending injection.

The pintle nozzle uses an inward opening valve principle requiring the use of leak-off lines/pipes to return fuel to the tank. The leakage takes place at the pintle valve to nozzle body clearance and tends to increase as these matched components age.

The shaping of the spray pattern emitted from a pintle nozzle depends on the pintle design. Throttling nozzles are designed to emit less fuel at the beginning of the injection pulse and may be identified by their conical shaped pintle valves.

Orifice Nozzles

As stated in the introduction to this chapter, most current truck and bus diesel engines use closed hydraulic injector nozzles. These nozzles are in almost all cases, of the multi-orifice type. Multi-orifice nozzles appear in truck engines in traditional, pump-line-nozzle configurations and they are an integral sub-component in EUIs and HEUIs: they are also used with EUP (electronic unit pump) fuelled engines. Orifice nozzles may have a single orifice or several orifices, almost always the latter. Multi-orifice nozzles are required in most high speed DI diesel engines as only these produce the necessary degree of atomization, droplets sized between 10 m–100 m. Atomized droplet sizing of 10 m or less can result in erratic ignition or no ignition at all. Atomized droplet sizing 100 m or more are too large to be completely combusted (oxidized) in the time dimension window of high-speed CI engines with rated speeds of 1800–2600 rpm. The time dimension window, within which combustion of the fuel must take place, is defined by engine rpm. If an engine is run at 2000 rpm, it takes exactly 0.030 seconds or 30 milliseconds for a single revolution of the engine: this means that a piston travels from TDC to 90 ATDC, more or less the period available for combustion, in 7.5 milliseconds. Top end engine speeds dictate the combustion or burn duration window and therefore the maximum atomized droplet sizing.

Orifice nozzles use an inward opening valve principle and that requires leak-off lines. The leakage is that occurring at the nozzle valve to nozzle body clearance and measured on the injector test bench as back leakage. NOP parameters in current engines range from 120–400 atms (1800–5880 psi) with peak system injection pressures ranging from 2–10 times the NOP value. Multi-orifice nozzles used in hydromechanical engines typically have NOP in the region of 200 atms or 3000 psi. The multi-orifice nozzles used in EUIs and HEUIs generally have NOP values closer to the higher range of the typical parameters, around 350 atms (5000 psi).

Action: The multi-orifice nozzle is usually dowel positioned in the injector assembly to ensure that the spray pattern is directed to a specific location of the combustion chamber or in the case of direct injection, to a specific location of the cylinder. Fuel from the injection pump element is delivered to the nozzle holder and then ducted in an annular recess in the upper nozzle valve body. Either single or multiple fuel ducts extend from the annular recess to the pressure chamber, meaning that the pressure here will be the same as that in the high-pressure pipe. The nozzle valve is loaded to a closed position on its seat, either directly by the injector spring or indirectly by spring pressure relayed to the nozzle valve by means of a spindle. This spring tension is adjustable by either a screw or by shims and will set the NOP value. When line pressure is driven upward by the injection pump, pressure will increase in the nozzle pressure chamber and this pressure is sufficient to overcome the mechanical force of the injector spring, the nozzle valve retracts permitting the fuel to flow past the seat into the sac and exit the nozzle orifice. At the center of the nozzle valve seat, a single duct connects to the nozzle sac, a spherical chamber into which the nozzle orifice is drilled. The sac hydraulically balances the fuel exiting the orifice, so that droplets start to exit from each orifice at approximately the same moment at the beginning of the fuel pulse. The injection pulse will continue for as long as the nozzle valve remains open; depending on the specific means used to generate injection pressures, maximum fuelling pulses can extend from 25–50 crank angle degrees.

The fuelling pulse ends when the metering/pumping element and the pressure in the nozzle assembly drops collapse the pressure to a value below NOP. The emitted droplet sizing from the nozzle orifice is factored by pressure and flow area. The flow area is represented by the sizing of the nozzle orifice and therefore remains constant. However, the pressure values vary considerably, extending from a value lower than NOP up to the peak pressure (the highest pressure attainable in a fuel injection system) value. As pressure increases, the droplet sizing will decrease. The more prolonged the injection pulse, (that is, the longer the duration of the effective pump stroke), the higher the circuit pressure at injection pump port opening (end of delivery) and the smaller the atomized droplet sizing. The reduction in droplet sizing that occurs as the fuelling pulse is extended is generally favorable for the complete combustion of the fuel, except for the short period of fuelling that takes place after pump effective stroke completion and before nozzle valve closure. As the pressure collapses in the pump element and, therefore, the injection circuit, the emitted droplet sizing will increase until the moment the nozzle valve actually closes, so it is generally desirable for this pressure collapse to occur rapidly.

Nozzle Differential Ratio

Nozzle differential ratio describes the geometric relationship between the sectional area of the nozzle seat and that of the pressure chamber or valve shank. Nozzle valves are opened by hydraulic pressure acting on the sectional area of the valve subject to the pressure chamber; when this overcomes the spring pressure that loads them on their seat, they retract. However, the instant the nozzle valve unseats, hydraulic pressure is permitted to act over the whole sectional area of the nozzle valve. The whole sectional area of the nozzle valve is the sectional area of both the seat and that of the pressure chamber combined.

When pressure rise begins in the injection line circuit prior to the opening of the nozzle valve, the hydraulic circuit which is being acted upon is closed (it is sealed at the nozzle seat). The instant the NOP value is achieved, the nozzle valve unseats and opens the circuit. This must result in a drop in line pressure. However, because at the moment the nozzle valve opens, the sectional area over which the hydraulic pressure acts is increased by that of the seat sectional area, less pressure is now required to hold the nozzle open. Nozzle differential ratio must be sufficient to prevent nozzle closure when this pressure drop occurs. After NOP, fuel passing around the seat fills the sac and pressure rise resumes because of the restriction represented by the minute sizing of the nozzle orifice. Nozzle differential ratio means that nozzle closure at the end of injection will occur at a value somewhat below the NOP value. It will also help define the specific residual line pressure (pressure that dead volume fuel is retained at in the high-pressure pipe) value.

More pressure is required to unseat a nozzle valve than that required holding it off its seat due to nozzle differential ratio.

Nozzle Holders/Injectors

Injectors are simply mounting devices for hydraulic nozzles. They come in many different shapes and sizes for a variety of reasons; for instance, long stem nozzles are easier to cool. Older hydraulic injectors tended to be of the high spring design. High spring injectors located the spring in the upper portion of the holder and relayed the spring tension to the nozzle value by means of a spindle. Spring tension was altered directly by an adjusting screw and setting the NOP value on the pop tester was a quick and easy procedure. A disadvantage of the high spring design was at NOP, the nozzle valve slammed open until it was mechanically prevented from further inboard travel and because of its high opening velocity, the spindle was driven into the spring which rebounded sufficiently to hammer the nozzle valve from its open position back into the seat, interrupting the pulse and affecting emitted droplet sizing. More recent hydraulic injector nozzles have tended to use a low spring design that eliminates the spindle and thus, reduced the mass of moving parts; this has minimized rebound interference of the injection pulse. However, low spring injectors generally use shims acting on the spring to define the NOP value, which extends the time required setting the NOP value on the test bench.

Pencil Type Injector Nozzles

Pencil injector nozzles are a type of multi-orifice injector nozzle and in fact, share common operating principles with a couple of small exceptions. They are seldom found in any current truck engine applications. Caterpillar has used them in some of their hydromechanical engines. The pencil nozzle assembly is cylindrical and has the approximate appearance of a pencil. Within the nozzle body, the nozzle valve extends through nearly the full height of the injector assembly and the injector spring acts directly on top of the nozzle valve shaft. Adjustment is by means of an adjusting screw and locknut located at the top of the assembly. Some pencil nozzles have no leak-off lines: fuel that bleeds by the nozzle valve during injection is accumulated in a chamber above the valve and pressure equalization with line pressure occurs after nozzle closure. Pencil nozzles are most often damaged in the process of removal and service. When making any adjustments, use the recommended mounting fixture and observe the specified torque procedure.

Valve Closes Orifice (VCO) Nozzles

Valve closes orifice nozzles have eliminated the sac. The function of the sac in the nozzle is to provide balanced fuel dispersal, which is especially important in keeping the ignition lag time dimension consistent. However, at the completion of the injection pulse the volume of fuel in the sac was essentially wasted fuel that added to HC emission. At the instant of nozzle closure, the sac and the nozzle orifice would contain fuel that would be vaporized due to the heat of combustion but at best, only partially combusted. Current injector nozzle designs have either substantially reduced the sac volume or eliminated it entirely. A true VCO nozzle has orifice extend directly from the seat. VCO nozzles are most often used on electronically managed injection systems that use high NOP values which can to some extent compensate for the compromising of the balanced fuel dispersal offered by nozzle sacs.

Nozzle Testing and Reconditioning

Injector nozzles are probably subjected to more abuse than any other engine component. They are exposed to temperature peaks of 14,000°C (25,500°F) outside and pressures up to 2000 atms internally. Manufacturer's service intervals are seldom respected and field diagnostics tend to be poor. The truck and bus diesel technician of today is seldom required to fully recondition nozzle assemblies due to the high cost of labor and service tooling, versus the relatively low cost of nozzle replacement. However, most manufacturers include the injectors in preventative maintenance schedules and all that is required in terms of equipment is a simple bench test fixture and ultrasonic bath. Servicing a set of injectors is a safe and easy procedure, but some precautions are required:

- Eye protection should be worn when working with or near high-pressure fluids.
- High pressure atomized fuel is extremely dangerous and no part of the body should be in danger of contacting it.
- Never touch the nozzle assembly on the pop tester when it is at any pressure above atmospheric—and remember that the whole injector assembly is under pressure.

Removal of Injectors from the Cylinder Head

The OEM recommended method of removing an injector from a cylinder head must be adhered to and failure to do so can result in failed injectors. Before attempting to remove injectors, thoroughly clean the surrounding area, remembering that any dirt around the injector bore can end up in the engine cylinder below it. Thin bodied pencil type nozzles are vulnerable to any side load force and the correct pullers should always be used. Additionally, whenever pencil type nozzles are removed from a cylinder head, they must be tested in a test fixture before reinstallation, because they are so easily damaged during the removal process. Many injector assemblies are flanged in which case, the hold down fasteners and clamps should first be removed and the injector levered out using and injector heal bar. The injector heal bar is 8 in.–12 in. (20–30 cm) in length to prevent the application of excessive force to the injector flange. Where cylindrical injectors are used, a slide hammer and puller nut that fits to the high-pressure inlet of the injector should be used to pull the injector. With certain cylindrical hydraulic injectors, the high-pressure delivery pipe fits to a recess in the injector through the cylinder head; the high-pressure pipe must be backed away from the cylinder head before attempting to remove the injector or both the injector and the seating nipple on the high-pressure pipe will be damaged. When removing injectors from cylinder heads, make a practice of removing the injector nozzle washer at the same time: these have an id less than the cylinder injector aperture id, so providing the piston is not at TDC, they can usually be removed either by inserting an O-ring pick or in cases of difficult to remove washers, by taking the tapered end of the injector heal bar and jamming it into the washer. It is not good practice to reuse injector washers. Both steel and copper washers harden in service. The copper washers can be annealed by heating to the point they shimmer green, then quenched with water. Steel washers are a poor reuse risk and no attempt should be made to anneal them.

When the injector has been removed, use plastic caps to seal the cylinder head injector bores, the injector inlet, and high-pressure pipe nipples. Ensure that a set of injectors is marked by cylinder number and properly protected in an injector tray: wrapping the injectors in shop rags will do if no injector tray is available.

Occasionally, an injector may seize in a cylinder head and defy any normal means used to remove it. Once the injector becomes damaged (for instance, the line threads that fit the injector to the puller are destroyed) the technician has no option but to remove the cylinder head: never risk damaging a cylinder head attempting to remove an injector. With the cylinder head removed, a seized injector can usually be removed with a punch and hammer.

Testing

OEM specifications and procedures should be consulted before the nozzle to be tested is placed in the nozzle test fixture. A typical procedure for testing nozzle assemblies would be as follows. Perform the following tests in sequence. At the first failure, cease following the sequence and proceed to the next section:

- Clean injector externally with a brass wire brush.
- Locate manufacturer's test specifications.
- Mount the injector in the bench test fixture. Build pressure slowly using the pump arm, watching for external leakage.
- Bench test NOP value and record. Use three discharge pulses and record the average value. Sticking or a variation in NOP value that exceeds 10 atms (150 psi) fails the nozzle.
- Test forward leakage by charging to 10 atms (150 psi) below the NOP value and holding the gauge pressure at that value while observing the nozzle. Any leakage evident at the tip orifice fails the nozzle.
- Check the back leakage factor by observing pressure drop from a value 10 atms (150 psi) below NOP. Pressure drop values should typically be in the range of 50–70 atms (700–1000 psi) over a 10 seconds test period. Pressure drop that exceeds the OEM specification indicates too little valve to body clearance (possibly caused by valve to body mismatch). A rapid pressure drop exceeding the OEM specification indicates excessive nozzle valve to body clearance, a condition usually caused by wear.
- Again, actuate the bench fixture pump arm and observe the nozzle spray pattern, checking for orifice irregularity. Ignoıe nozzle chatter (rapid pulsing of the nozzle valve): this can be regarded as a test bench phenomenon due to slow rate of pressure rise. In some modern injectors, the intensity of chatter noise is reduced due to the use of double seats.

Reconditioning

The function of the assessment procedure is to determine whether the nozzle should be reconditioned or placed back in service. Reconditioning in the service shop of today seldom means regrinding of the valve nozzle seat or reaming nozzle orifice due to high labor and tooling costs in comparison with the cost of nozzle replacement. The following simple procedure probably summarizes what most operations interpret as reconditioning:

- A nozzle that fails the inspection procedure should first be removed from the injector. Disassemble the injector and remove the nozzle, separating the nozzle valve from the nozzle body. Ensure that the nozzle valves remain matched with the nozzle bodies they are removed from because they are not interchangeable.
- Prepare the nozzle valve and bodies for ultrasonic cleaning. An ultrasonic bath filled preferably with reasonably clean soapy water should be used: in other words, change the cleaning solution regularly. Alternatively diesel test oil can be used, but it is not as effective. A five-minute, ultrasonic bath is generally sufficient followed by dehumidification in either a dehumidifier or an oven. If a dehumidifier is not

available, fuel test oil rather than soapy water should be used as the ultrasonic soaking medium. Remember that lapped components are not interchangeable so if cleaning multiple nozzle assemblies do not mismatch components. CAUTION! Ultrasonic baths emit sound waves that are outside the frequencies detected by the human ear and have been known to trigger seizures in those susceptible to epilepsy. It is good practice to vacate a room in which ultrasonic cleaning is being performed although it represents to most people nothing more than an irritating buzzing.
- Test or replace the injector spring. Reassemble the injector assembly using diesel fuel as an assembly lubricant. Never use engine oil or any grease base lubricant.

Generally, set the NOP value at 10 atms (150 psi) above the recommended specification if no parts have been replaced. Add another 5 atms (75 psi) if using a new spring and a further 5 atms (75 psi) if using a new nozzle assembly. (After 100 engine hours, the nozzle will function at the specified NOP value.) In other words, when reassembling an injector with a new nozzle assembly and a new spring, the NOP would be set 20 atms (300 psi) above the recommended specification. Always consult the appropriate service literature for the NOP setting procedure. Use the following as a guide:

- Screw type: back off lock nut, then turn adjusting screw CW or CCW to either increase or decrease NOP.
- Shim type: add or subtract shims acting on the injector spring.
- Check forward leakage.
- Check back leakage.
- Check back leakage pipe restriction.
- Observe spray pattern and check for nozzle orifice irregularity, preferably using a template.
- A primary function of the nozzle washer (spacer) is to ensure optimum droplet penetration into the engine cylinder. It is not a mere joint gasket. Spacer thickness is critical and determines the nozzle protrusion dimension. Use of an incorrectly sized washer, two washers, or no washer can result in rapid engine failures.

Reinstallation of Injectors

The injector bore should be cleaned before installation using a bore reamer if necessary to remove carbon deposits. Blow out the injector bore using an air nozzle. Most diesel engine OEMs prefer that injectors be installed dry: lubricants such as Never-Seize can hinder the ability of the injector to transfer heat to the cylinder head. It is good practice to turn the engine over and pump fuel through the high-pressure pipe prior to connecting them to the injector: this will purge the pipe of air and possibly debris that may have intruded into the line while it was disassembled. Always torque the nut on the high-pressure pipe with a line socket: failure to this can result in damage to both the line nut and the nipple and seat.

Task F1.11 Inspect, adjust, repair/replace smoke limiters (air/fuel ratio controls).

Aneroid

By definition, an aneroid is a low-pressure sensing device. In application, it is used on a turbocharged diesel engine to measure manifold boost and limit fueling until the boost pressure achieves a predetermined value.

Such devices are known variously as puff limiters, turbo boost sensors, AFC valves, and smoke limiters and all seek to accomplish the same objective. They typically consist of a manifold within which is a diaphragm; boost air is piped from the intake manifold to act on the diaphragm. Such devices are used on most current boosted engines (turbocharged engines).

When an aneroid is used on an in-line, port-helix metering pump, it is usually a mechanism consisting of a manifold, spring, and control rod. The manifold is fitted with a port and a steel line connects it directly to the engine intake manifold. In this way, boost pressure is delivered to the aneroid manifold and where it will act directly on the diaphragm within. Attached to the diaphragm is a linkage connected either directly or indirectly to the fuel control mechanism (rack): a spring loads the diaphragm to a closed position which will limit fueling by preventing the rack from moving into the full fuel position. When manifold boost acting on the diaphragm is sufficient to overcome the spring pressure, it acts on the linkage permitting the rack full travel and thus maximum fueling. Such systems are easily and commonly shorted out by operators in the mistaken belief that aneroid systems reduce engine power. The emission of a "puff" of smoke from the exhaust stack at each gear shift point is an indication that someone has tampered with the aneroid/boost sensing mechanism.

Altitude Compensator

An altitude compensator contains a barometric capsule that measures barometric pressure. Based on this pressure, it down rates engine power at higher altitudes to prevent overfueling. They are required when running at higher altitudes because the oxygen density in the air charge decreases with an increase in altitude and unless the fuel system is aware of this, it effectively overfuels the engine. The critical altitude at which some measure of injected fuel quantity deration is required to accord with today's low emission engines is 1,000 ft. (303 meters) in altitudes. Older engines with altitude compensators usually did not derate the fuel system until an altitude of at least 2,000-ft. (606 meters)

Task F1.12 Inspect, reinstall/replace high-pressure injection lines, fittings, and seals.

To remove high-pressure fuel lines:

- Remove the low-pressure fuel lines.
- Disconnect the fuel injection lines at the injectors. Cap each injection line and injector. The line is removed to prevent the entry of dirt into the system.
- Disconnect the fuel lines at the fuel injection pump. Cap each line and pumping element as the line is removed to prevent the entry of dirt in the system.
- Detach the three brackets that hold the fuel injection line assembly to the intake manifold.
- Carefully wiggle the fuel injection lines out from under the intake manifold and remove them from the engine. Start from the back and move toward the front of the engine.

Once removed from the engine, individual fuel lines can be removed from the assembly for replacement. Keep in mind when ordering parts that each line is different and cannot be interchanged; therefore, order lines individually for the particular cylinder of that the line.

To install high-pressure fuel lines:

- Make sure that the clamps holding the lines together are tight. High-pressure fuel lines must be installed as a set. It is much easier than installing them individually.
- Wiggle the high-pressure fuel injection lines into position on the engine. Start from the back and go toward the front of the engine.
- Install the injection lines onto the injection pump and tighten the nuts. Tighten the brackets holding the high-pressure fuel lines to the intake manifold.
- Connect the high-pressure fuel injection lines to the fuel injectors one at a time starting with cylinder No. 1. Do not remove the caps from the fuel lines or components until they are to be connected. This will help prevent the entry of dirt into the system.
- Tighten the fuel injection line to the injector nuts.
- Install the low-pressure fuel lines and filter assembly back on the engine. Tighten the nuts to specifications.
- Connect the battery ground cables to both batteries.
- Run the engine and check for leaks.
- If necessary, purge the high-pressure fuel lines of air by loosening the connector one-half to one turn and cranking the engine until solid fuel, free from bubbles, sprays from the connection.

Injector leak off line removal:

- Remove banjo bolts that hold the leak off line to the fuel injector. Start at No. 1 injector and work backward.
- Store banjo bolts in a clean place to avoid contamination. Discard washers. New washers must be used for re-assembly.
- Disconnect the leak off line at fuel box and remove it from the engine.

Note: Reverse procedure for installation.

Task F1.13 Inspect, repair, reinstall/replace low-pressure fuel lines, cooling plate, fittings, and seals.

Low-pressure fuel injection line removal:

Before removing any fuel lines, clean the exterior of the fuel lines with fuel oil or solvent to prevent the entry of dirt into the fuel system when the fuel lines are removed. Blow-dry with compressed air.

- Disconnect the battery ground cables from all batteries.
- Disconnect and remove the low-pressure fuel line assembly, and cap the lines. Inspect the rubber spacers at the ends of the lines for distortion. Replace as required.

Installation of low-pressure fuel injection lines:

- Install the low-pressure fuel lines back on the engine. Make sure the tubes have fully seated into their correct fittings before tightening retaining nuts to specification. Replace any seals previously discarded.
- Secure lines to prevent damage and kinking. Some modern diesel engines use newer automotive design quick connectors. These connectors snap and clip together providing a good seal when properly connected. To disconnect and reconnect the lines, a special insert tool must be used. Forcing lines apart can cause damage to the connections. When installing the lines ensure adequate room exists to remove the insertion tool.
- Connect the battery ground cable to all batteries.

Task F1.14 Inspect, test, adjust, repair/replace engine governor systems.

The mechanical governor is an integral part of the pump or fuel rack that controls the injector. It is either belt or gear driven from the engine. The word "hunting" refers to an unstable governor causing an unstable engine. This is generally caused by the governor over-correcting for load. This usually happens when a no-load condition exists, or the governor oil is low. Although some hydraulic governors incorporate a droop function, which effectively causes the governor to become sluggish, reducing over-compensation during load changes.

Task F1.15 Inspect, test, adjust, repair/replace engine fuel shut-down devices and controls.

Task F1.16 Inspect, test, adjust, repair/replace safety shut-down devices, circuits, and sensors.

The fuel shutoff solenoid is comprised of two coils within one housing. A powerful "pull" coil is used to pull the fuel shutoff lever to the "open" position. The pull coil is fed from the switched side of the starter motor relay, so it is energized whenever the ignition switch is in the Start position. At the same time, a much less powerful "hold" coil is fed from a "Start/Run" circuit. When the ignition switch is released to the Run position, the pull coil drops out and the hold coil is left to keep the fuel shutoff lever in the "open" position. Because the pull coils draw a considerable amount of current while it is energized, do not allow it to remain energized for longer than 30 seconds, or it will fall open. In instances of a no-start condition where the engine turns over normally and all correct starting procedures are being followed, it is a good idea to take a look at the fuel shutoff solenoid to make sure it is moving the lever when the key is turned to the Start and Off positions. In some cases, it may be necessary to adjust the travel of the arm on the fuel solenoid. To adjust the fuel shutoff solenoid, perform the following procedure:

- With the key in the Off position, pull the ball adjuster off the rack control lever. Manually bottom the solenoid plunger and check to ensure manufacturer's specifications are met for travel before the lever bottoms out. If the lever bottoms before the solenoid, use an adjuster to set the proper clearance. Attach the ball adjuster to the rack control lever.
- Turn the stop screw in, so that the rack control lever will not contact the external stop when the key is in the Off position. Turn the key to the Off position. The rack control should go all the way to the internal pump stop.
- Adjust the external stop screw to the point where it touches the rack control lever. Then, turn the screw half to one turn beyond this point, so that travel of the rack control is stopped completely by the external stop.

To remove the fuel shutoff solenoid:

- Disconnect the wires, which attach to the bottom of the fuel shutoff solenoid.
- Pull the ball adjuster off of the rack control lever.
- Remove the cap screws, holding the fuel shutoff solenoid to the side of the fuel injection pump. Remove the fuel shutoff solenoid from the engine.

To install the fuel shutoff solenoid:

- Attach the fuel shutoff solenoid to the side of the fuel injection pump with cap screws. Tighten the cap screws to manufacturer's specification.
- Connect a ball adjuster to the rack control lever.
- Attach wires to the bottom of the fuel shutoff solenoid. The terminal sizes may be different but wires should not go on the wrong terminals.
- Adjust the fuel shutoff solenoid as previously outlined.

2. Electronic Components (10 Questions)

Task F2.1 Check and record engine electronic diagnostic codes and trip/operational data; clear codes; determine needed repairs.

Modern engines are controlled and monitored by electronic modules. Some of these modules are capable of storing PID error codes against engine parameter identifications. To retrieve these codes, the technician connects a diagnostic test set to a diagnostic plug. Some vehicles will display a limited readout of PIDs by blinking an instrument panel indicator in a two-digit code called a flash code. Diagnostic test sets also have a menu driven system and module check that can be run. To test the engine, once the problems have been analyzed and corrected, the PID codes need to be cleared. Diagnostic test sets can be used to clear these codes. Another approach is to disconnect the battery.

Electronic service tools capable of reading ECM data are connected to the on-board electronics by means of a SAE (Society of Automotive Engineers)/ATA (American Trucking Association) J1584/J1708/1939 6 pin Deutch connector in all current systems. This data connector is used by all the truck engine electronics OEMs and is most often referred to as an ATA connector. This common connector and the adherence by the engine electronics OEMs to SAE software protocols enables proprietary software of one manufacturer to at least read the parameters and conditions of their competitors: this means that if a Navistar powered truck has an electronic failure in a location where the only service dealer is DDC, some basic problem diagnosis can be undertaken using the DDC electronic diagnostic equipment.

Output Circuit

Output circuit devices are components that are controlled by the ECM: they are the set of components that effect the results of the programmed software and the processing computations. Any component that is switched or can be controlled by the ECM can be called an output.

There are six categories of electronic service tools as follows:

1. Digital Multimeter—The digital multimeter (DMM) is an essential diagnostic instrument that should be part of every truck technician's tool chest. You use it to troubleshoot electronic circuits and test system components.

DMM Operating Principles

In any electrical circuit voltage, current flow, and resistance can be calculated using Ohm's Law. A DMM makes use of Ohm's Law to measure and display values in an electrical circuit. A typical DMM will have the following selection options chosen by rotating the selector to one of the following:

off	Shuts down DMM
V~	Enables AC voltage readings
V–	Enables DC voltage readings
mV-	Enables low pressure DC voltage readings
Ω	Enabales component or circuit resistance readings
α	Enaables continuity (a circuit capable of being closed) testing Identifies an open/closed circuit
A~	Checks current flow (amperage) in an AC circuit
A–	Checks current flow (amperage) in a DC circuit

Measuring Voltage

Checking circuit supply voltage is usually one of the first steps in troubleshooting. This would be performed in a vehicle DC circuit by selecting the V setting and checking for voltge present or high/low voltage values. Most electronic equipment is powered by DC voltage. For example, home electronic apparatus such as computers, televisions and stereos use rectifiers to convert main AC voltage to DC voltgae.

The waveforms produced by AC voltages can either be sinusoidal (sine waves) or nonsinusoidal (sawtooth, square, ripple). A DMM will display the **root mean square** (RMS) value of these voltage waveforms. The RMS value is the effective or equivalent DC value of the AC voltage. Meters described as "average responding" give accurate RMS readings only if the AC voltage signal is a pure sine wave; they will not accurately measure nonsinusoidal signals. DMMs described as "true-RMS" will measure the correct RMS value regardless of waveform and should be used for nonsinusoidal signals.

A DMM's ability to measure voltage can be limited by the signal frequency. The DMM specifications for AC voltage and current will identify the frequency range the instrument can accurately measure.

Voltage measurements determine:

1. Source voltage
2. Voltage drop
3. Voltage imbalance
4. Ripple voltage
5. Sensor voltages

Measuring Resistance

Most DMMs will measure resistance values as low as 0.1 ohm and some will measure high resistance values up to 300 MΩ (mega ohms). Infinite resistance or resistance greater than the instrument can measure is indicated as "OL" on the display. For instance an open circuit (one in which there is no path for current flow) would read "OL" on the display.

Resistance and continuity measurements should be made on open circuits only. Using the resistance or continuity settings to check a circuit or component that is energized will result in damage to the test instrument. Some DMMs are protected against such "accidental" abuse and the extent of damage will depend on the model. For accurate low resistance measurement, test lead resistance, typically between 0.2 Ω–0.5 Ω depending on quality and length, must be subtracted from the display reading. Test lead resistance should NEVER exceed 1 ohm.

If a DMM supplies less than 0.3V DC test voltage for measuring resistance, it is capable of testing resistors isolated in a circuit by diodes or semiconductor junctions, meaning that they do not have to be removed from the circuit board.

Resistance measurements determine:

- Resistance of a load
- Resistance of conductors
- Value of a resistor
- Operation of variable resistors.

Continuity is a quick resistance check that distinguishes between an open and a closed circuit. Most DMMs have audible continuity beepers that beep when they detect a closed circuit, permitting the test to be performed without looking at the meter display. The actual level of resistance required to tirgger the beeper varies from model to model.

Continuity tests determine:

- Fuse integrity
- Open or shorted conductors
- Switch operation
- Circuit paths

Diode Testing

A diode is an electronic switch that can conduct electricity in one direction while blocking current flow in the opposite direction. It is commonly enclosed in a glass cylinder; a dark band identifies the cathode or blocking terminal. Current flows when the anode is more positive than the cathode. Additionally, a diode will not conduct until the forward voltage pressure reaches a certain value (0.3V in a silicon diode). Some meters have a diode test mode. When testing a diode with the DMM in this mode, 0.6V is delivered through the device to indicate continuity; reversing the test leads should indicate an open circuit in a properly functioning diode. If both readings indicate an open circuit condition, the diode is open. If both readings indicate continuity, the diode is shorted.

Measuring Current

Current measurements are made in series, unlike voltage and resistance readings which are made in parallel. The test leads are plugged into a separate set of input jacks and the current to be measured flows through the meter.

Current measurements determine:

- Circuit overloads
- Circuit operating current
- Current in different branches of a circuit

When the test leads are plugged into the current input jacks and are used to measure voltage, this causes a current shunt—direct short across the source voltage through a low value resistor inside the DMM. A high current flows through the meter and, if not adequately protected, both the meter and the circuit can be damaged.

A DMM should have current input fuse protection of high enough capacity for the circuit being tested. Ths is mainly of importance when working with high pressure (220V+) circuits.

2. On-board diagnostic lights—Blink or flash codes (OEMs use both terms) are an on-board means of troubleshooting using a dash or ECM mounted electronic malfunction light or CEL (check engine light). Normally only active codes (one indicating a malfunction at the time of reading) can be read in truck (engine) ECMs but there are exceptions.

3. Scanners—These are read-only tools capable of reading active and historic or inactive (logged but not currently indicating a malfunction) codes and sometimes system parameters but little else. They are obsolete as a truck/bus diesel engine diagnostic tool.

4. SCAN TOOLS generic reader/programmers—Microcomputer-based electronic service tools that are designed to read and reprogram all proprietary systems in conjunction with the appropriate software cartridge. (ProLink 9000 has become the industry standard and this key diagnostic instrument is examined in some detail later on in this section.

5. Proprietary reader/programmers—These are usually PC-based test instruments packaged by the OEM for use exclusively on their system. They have the advantage of offering an optimum degree of user friendliness and the disadvantage of being system specific and high in cost. Examples are:

- Cummins: Compulink, Echeck
- Cat: ECAP
- DDC: DDR Programming Station

6. Personal computer (PC)—Most OEMs (DDC, CAT, Cummins, and MACK) have made the generic PC and its operating systems (DOS/MS Windows 3.1/Windows 95, Windows 98) their diagnostic and programming tool of choice; the remainder are expected to follow. This makes sense.

PCs are cheap, easily upgradable and have vast computing power when compared with proprietary reader/programmers. OEMs can upgrade/update software more easily and the PC is not confined to the limitations of all the other existing electronic troubleshooting tools. PCs are connected to vehicle ECM through the SAE/ATA J1708/1939 6 pin connector: a serial link or interface module may also be required, depending on the system.

Truck engine OEMs have chosen to use DOS and MS Windows driven programs and offer comprehensive courses on their own management systems which usually include a thorough orientation of PCs. PCs are addressed in some detail in a preceding chapter of this text.

SAE/ATA J1587/J1708/J1939 codes and protocols—The following is a partial listing of SAE/ATA codes that have been adopted by all the North American truck engine/electronics OEMs. Generally, SAE J 1587 covered common software protocols in electronic systems, SAE J 1708 covered common hardware protocols, and the more recent SAE J 1939 covers both hardware and software protocols. The acceptance and widespread usage of these protocols enables the interfacing of electronic systems manufactured by different OEMs on truck and bus chassis, plus provides any manufacturer's software to at least read other OEMs electronic systems. The truck technician who works with multiple OEM systems may find it easier to work using these codes rather than use the proprietary codes.

The acronym MID is used to describe a major vehicle electronic system, usually with independent processing capability.

Message Identifier (MID)	Description
128	Engine Controller
130	Transmission
136	Brakes—Antilock/Traction Control
137–139	Brakes—Antilock, Trailer #1, #2, #3
140	Instrument Cluster
141	Trip Recorder
142	Vehicle Management System
143	Fuel System
162	Vehicle Navigation
163	Vehicle Security
165	Communication Unit—Ground
171	Driver Information System
178	Vehicle Sensors to Data Converter
181	Communication Unit—Satellite

The acronym PID is used to code components within an electronic subsystem.

Parameter Identifier (PID)	Description
65	Service Brake Switch
68	Torque Limiting Factor
70	Parking Brake Switch
71	Idle Shutdown Timer Status
74	Maximum Road Speed Limit
81	Particulate Trap Inlet Pressure
83	Road Speed Limit Status
84	Road Speed
85	Cruise Control Status
89	PTO Status
91	Percent Accelerator Pedal Position
92	Percent Engine Load
93	Output Torque
94	Fuel Delivery Pressure
98	Engine Oil Level
100	Engine Oil Pressure
101	Crankcase Pressure
102	Manifold Boost Pressure
105	Intake Manifold Temperature
108	Barometric Pressure
110	Engine Coolant Temperature
111	Coolant Level
113	Governor Droop
121	Engine Retarder Status
156	Injector Timing Rail Pressure
157	Injector Metering Rail Pressure
164	Injection Control Pressure
167	Charging Voltage
168	Battery Voltage
171	Ambient Air Temperature
173	Exhaust Gas Temperature
174	Fuel Temperature
175	Engine Oil Temperature
182	Trip Fuel
183	Fuel Rate
184	Instantaneous MPG
185	Average MPG
190	Engine Speed

The acronym SID is used to identify the major subsystems of an electronic circuit.

Subsystem Identifiers (SID) common to all MIDs	Description
242	Cruise Control Resume Switch
243	Cruise Control Set Switch
244	Cruise Control Enable Switch
245	Clutch Pedal Switch
248	Proprietary Data Link
250	SAE J1708 (J1587) 1939 Data Link

Subsystem Identifiers for MID 128, 143	Description
01–16	Injector Cyl #1 thru #16
17	Fuel Shutoff Valve
18	Fuel Control Valve
19	Throttle By-pass Valve
20	Timing Actuator
21	Engine Position Sensor

22	Timing Sensor
23	Rack Actuator
24	Rack Position Sensor
29	External Fuel Command Input

Subsystem Identifiers for MID 130	Description
1–6	C1–C6 Solenoid Valves
7	Lockup Solenoid Valve
8	Forward Solenoid Valve
9	Low Signal Solenoid Valve
10	Retarder Enable Solenoid Valve
11	Retarder Modulation Solenoid Valve
12	Retarder Response Solenoid Valve
13	Differential Lockout Solenoid Valve
14	Engine/Transmission Match
15	Retarder Modulation Request Sensor
16	Neutral Start Output
17	Turbine Speed Sensor
18	Primary Shift Selector
19	Secondary Shift Selector
20	Special Function Inputs
21–26	C1–C6 Clutch Pressure Indicators
27	Lockup Clutch Pressure Indicator
28	Forward Range Pressure Indicator
29	Neutral Range Pressure Indicator
30	Reverse Range Pressure Indicator
31	Retarder Response System Pressure Indicator
32	Differential Lock Clutch Pressure Indicator
33	Multiple Pressure Indicators

Subsystem Identifiers for MID 136–139	Description
1	Wheel Sensor ABS Axle 1 Left
2	Wheel Sensor ABS Axle 1 Right
3	Wheel Sensor ABS Axle 2 Left
4	Wheel Sensor ABS Axle 2 Right
5	Wheel Sensor ABS Axle 3 Left
6	Wheel Sensor ABS Axle 3 Right
7	Pressure Modulation Valve ABS Axle 1 Left
8	Pressure Modulation Valve ABS Axle 1 Right
9	Pressure Modulation Valve ABS Axle 2 Left
10	Pressure Modulation Valve ABS Axle 2 Right
11	Pressure Modulation Valve ABS Axle 3 Left
12	Pressure Modulation Valve ABS Axle 3 Right
13	Retarder Control Relay
14	Relay Diagonal 1
15	Relay Diagonal 2
16	Mode Switch—ABS
17	Mode Switch—Traction Control
18	DIF 1—Traction Control Valve
19	DIF 2—Traction Control Valve
22	Speed Signal Input
23	Warning Light Bulb
24	Traction Control Light Bulb
25	Wheel Sensor ABS Axle 1 Average
26	Wheel Sensor ABS Axle 2 Average
27	Wheel Sensor ABS Axle 3 Average
28	Pressure Modulator, Drive Axle Relay Valve

29	Pressure Transducer, Drive Axle Relay Valve
30	Master Control Relay
Subsystem Identifiers for MID 162	Description
1	Dead Reckoning Unit
2	Loran Receiver
3	Global Positioning System (GPS)
4	Integrated Navigation Unit

Failure Mode Identifiers (FMIs) are indicated whenever an active or historic code is read using ProLink or a PC.

Failure Mode Identifier (FMI)	Description
0	Data valid but above normal operating range
1	Data valid but below normal operating range
2	Data erratic, intermittent, or incorrect
3	Voltage above normal or shorted high
4	Voltage below normal or shorted low
5	Current below normal or open circuit
6	Current above normal or grounded circuit
7	Mechanical system not responding properly
8	Abnormal frequency, pulse width, or period
9	Abnormal update rate
10	Abnormal rate of change
11	Failure mode not identifiable
12	Bad intelligent device or component
13	Out of calibration
14	Special instructions

Task F2.2 Inspect, adjust, repair/replace electronic throttle and PTO control devices, circuits, and sensors.

One example of electronic control introduction into mechanical type fuel pumps is the addition of a throttle position sensor. This is intended to supply the power control module (PCM) with the position of the throttle so the PCM can regulate engine performance. To adjust this sensor it is necessary to connect a diagnostic tool up to the vehicle and clear all faults. Disconnect the linkages to the throttle and turn on the key, but do not start the engine during the diagnostic routine. The diagnostic tool will sound a beeping indication of the position of the sensor. If the setting is too high, the beeps will be fast; if it is too low, the beeps will be slow. An ideal setting will have a steady tone. To make these adjustments it may be necessary to remove the screws holding the FIPL sensor to the mounting bracket. Do not try to adjust the maximum throttle travel screw.

The throttle position sensor (TPS) is a variable resistor that signals the ECM the degree of throttle opening. Some engine manufacturers refer to the TPS as a "Williams Pedal." The sensor is connected to a 5-volt reference and has the highest resistance at closed throttle. At wide-open throttle, the resistance is lowest and output to the ECM will be near 5 volts.

- Connect the scan tool to the data link connector and select throttle position sensor.
- CManually close, the throttle outlet scan tool; it should indicate a low voltage.
- CManually open the throttle; the scan tool should indicate a high voltage.

To check the fuel injector pump lever (FIPL) sensor for proper operation and to make any adjustments, the engine must not be running (turned off). The throttle must be held to the floor during Key On Engine Off (KOEO) on-board diagnostics until the codes have begun to issue from the scan tool.

1. Perform KOEO on-board diagnostics and wait for all the service codes to be issued. If any error codes are present see the proper section in the "Engine/Emission Manual," correct as required, and restart this procedure.

2. After the last service code has been issued, press the Overdrive Cancel Switch (OCS). This will initiate the FIPL adjustment mode and allow the scan tool to be used as an "audible guide" in setting the FIPL sensor. Note: The scan tool remains in the adjustment mode for only 10 minutes; steps 3–6 must be completed within this time.
3. Remove the throttle cable from the throttle lever on the right side of the fuel injection pump.
4. Remove the throttle return spring. Install one end of the spring over the throttle lever ball stud and the other end over the throttle cable support bracket.
5. Insert a gauge block between the gauge boss and the maximum throttle travels screw. The throttle return spring, as repositioned in the above step, will hold the throttle lever open against the gauge block.
6. The FIPL sensor bracket is permanently attached to the pump with tamper proof screws. Movement of the bracket is not intended as a means for adjustment. If required, adjustment of the FIPL sensor may be accomplished by using the clearance between the sensor to bracket screws and the sensor.

To adjust, loosen the two screws that attach the FIPL sensor to the mounting bracket. Rotate the FIPL sensor until a steady tone is heard from the scan tool. If the setting is too low, the scan tool will issue a slow beep. If the setting is too high, the scan tool will issue a fast beep. Once a steady tone is heard, tighten the attaching screws to 8–10.5 Nm (75–90 in.-lb.). If the FIPL sensor cannot be adjusted to obtain a steady tone, replace the FIPL sensor and repeat this procedure from step 1.

Remove the gauge block. Cycle the throttle lever from idle to wide-open throttle five times. Reinsert the gauge block to verify the setting. If the tone is not steady, then readjustment is necessary. Repeat this procedure from step 6.

It is important that the FIPL sensor be set to the correct position (a steady tone). If the sensor setting is too low (a slow beep) the shifts will be firmer and more delayed than normal; if it is too high, (a fast beep) the shift will be smoother and earlier than normal.

If the 10-minute time limit has been exceeded and the scan tool tone has shut off, repeat this procedure from step 1. Remove the gauge block. Reattach the throttle return spring and throttle cable in the proper locations. Start the engine. Check throttle operation and transmission shift scheduling and quality.

Task F2.3 Perform on-engine inspections, tests, and adjustments on distributor-type injection pump electronic controls.

Task F2.4 Perform on-engine inspections, tests, and adjustments on in-line type injection pump electronic controls.

Electronic In-Line Pump Timing

You use this pump timing method with later generation hydromechanical and electronically managed, port-helix metering injection pumps. One checks the static timing value with an electronic timing tool that senses the positioning of a raised notch located on the pump camshaft.

Injection pump to engine timing is critical. One should check timing when reinstalling a pump to the engine and as a step in performance complaint troubleshooting. Pump timing may be checked using a diesel engine timing light: This consists of a transducer that clamps to the high-pressure pipe and signals the light when it senses the line pressure surge that occurs when the delivery valve opens. The test is not an accurate one and the values read on the timing light should not be confused with the pump manufacturer's port closure specification as the pulse used to trigger the light occurs after port closure. The test has some validity as a comparative test (using data obtained from other engines in the same series, timed at the same port closure value) and is a good means of verifying the operation of timing advance mechanisms.

Static Timing a Bosch Elctronically Controlled Injection Pump

The following procedure sequences the static timing of an electronically controlled injection pump. It references the procedure required to check and correct the timing of a Bosch PE 7100 or PE 8500 injection to a Mack Trucks E-7; V-MAC I or II managed engine. The procedure is outlined here as an example of how a current port-helix injection pump to engine. It is further important to note that Mack have used both #1 and #6 cylinders to time V-MAC injection pumps to the engine: early versions were timed on #6 cylinder, while later versions are timed on #1.

Ensure that the exterior of the injection pump and engine are clean. Remove the TEM (timing event marker) sensor from the rack actuator housing. Use a fixed timing sensor (J 37077) and ensure that the probe terminals are clean. Install the probe into the TEM port and ensure that it is securely locked into place.

Next ground the fixed timing tool ground clamp to a good ground on the engine. Mack use both front and rear timing indicators and the static timing procedure is different for each. The procedure outlined here will be based on that used for a rear located timing indicator.

Remove the flywheel housing timing window on the flywheel housing and locate the engine calibration scale and timing marker. Remove the plastic plug in the front of the flywheel housing and insert a toothed engine barring socket (J 38587) and ensure that it engages with the ring gear teeth on the flywheel. This barring tool enables the engine to be rotated with ease using a 1/2 in. drive ratchet of 12 in. length.

Rotate the engine using the barring tool to position it somewhere around 45 degrees BTDC. The fixed timing sensor tool is designed to determine the exact location of port closure in the injection pump by reading the timing notch located on the reluctor wheel in the rack actuator housing. There are two lights on the fixed timing sensor tool marked A and B. Each will illuminate for approximately 10 crank angle degrees on either side of the notch off the reluctor wheel. If the starting point is 45 degrees BTDC on the cylinder being timed and the engine is rotated in correct direction of rotation (clockwise), the A light will illuminate at 10 degrees before port closure. The objective is to stop at the precise moment that both A and B lights illuminate. This procedure is designed to read the actual port closure value within a 1/2 crank angle degree of accuracy. If the engine is barred past the point both lights illuminate so that the B light only is lit, the engine should be backed up and the procedure repeated: never back the engine up to locate pump port closure due to the engine time gear train backlash variables.

Next, check the engine calibration scale and timing marker. The reading should be within a 1/2 crank angle degree of specification with both lights on the fixed timing sensor tool illuminated. If the reading is within the required specification, remove the barring and timing tools and reinstall the TEM sensor according to Mack technical service literature.

If the reading is out of specification, verify that the timing marker is correctly positioned, that is correctly reading TDC when the engine is at true TDC. NEVER assume that the pump is incorrectly timed to an engine without first checking the timing marker. To correct static timing, proceed as follows:

Position the engine at the correct static timing location ensuring that it barred clockwise to position it there. Remove the injection pump drive gear access cover. Loosen the pump drive gear capscrew and fit a hub rotation tool onto 2 of the capscrew: this will permit the timing gear hub to be rotated independently of the timing gear. Using the hub rotation tool, turn the timing gear hub counter-clockwise until the capscrew lock against the slotted holes in the timing gear: at this point, neither of the timing lights should be illuminated.

Next, carefully rotate the gear hub clockwise until both the A and B lights are illuminated. Torque all four timing gear capscrew to specification. During the torquing procedure, it is normal for one of the lights to extinguish. Lubricate the timing gear access cover and torque to specification. Remove the barring and timing tools and reinstall the TEM sensor using the appropriate Mack procedure.

Task F2.5 Perform on-engine inspections, tests, and adjustments on PT-type injection pump electronic controls.

Cummins introduced CELECT (Cummins electronic engine control), their full authority electronic management system in 1991 on L10 and N14 engines, and later on their M11 engine when it was introduced in 1994. CELECT Plus electronics were released into the marketplace in 1996 and are used to manage current M11 Plus and N14 Plus engines.

CELECT Plus is a full authority, computerized engine management system that uses cam actuated, electronically controlled CELECT injectors which can be likened to the EUIs on other full authority systems. In fact, CELECT Plus has much in common with its competitor's full authority management systems and to properly understand this section, the reader should have a thorough understanding of hydromechanical fuel systems and basic vehicle computer operation covered elsewhere in this text.

Task F2.6 Perform on-engine inspections, tests, and adjustments on hydraulic electronic unit injector (common rail or rail pressure control) system electronic controls.

A typical electronic unit injector system of today has many improvements over older mechanical ones. In the Navistar/Caterpillar Hydraulic Electronic Unit Injector or HEUI system, fuel is not only pumped to the filter but with the use of a tandem pump; it can also be pumped from the filter. To develop necessary fuel injection pressure, diesel injectors in the application are powered by lubricating oil, which is pressurized by a special pump in the engine valley. The pump output pressure ranges from 450 psi to 3,000 psi (3,102 to 20,685 kPa). This oil pressure is controlled by the powertrain control module (PCM) with a spill valve called the injector pressure regulator. The high pressure lubricating oil is delivered to oil rails in the cylinder heads. An injection control pressure sensor mounted on one of the oil rails sends an analog voltage signal (0.5v to 5.0v) to the PCM for feedback control of the oil pressure. Fuel is drawn from the fuel tank through the primary filter by the diaphragm section of the tandem fuel pump. Pressurized fuel (approximately 28 kPa (4 psi) is supplied to the secondary filter and returned to the second stage of the tandem fuel pump. The piston-actuated second stage of the tandem fuel pump supplies 40–60 psi (276–414 kPa) of fuel to the rear of each cylinder head, where it flows to a fuel rail machined in each cylinder head. Drilled openings in the cylinder head route the fuel to the plunger area of the fuel injector, which can pressurize the fuel to 21,000 psi (144,795 kPa or 144 MPa) for delivery to the combustion chamber via a conventional nozzle/valve tip arrangement. The high pressure lubricating oil flows from the oil rails into an amplifier piston located in the injector. Oil entry to and exit from the amplifier piston is controlled by a solenoid-operated poppet valve. The fuel plunger is located in the injector and is driven by the amplifier piston. The fuel plunger injects fuel into the combustion chamber at pressures of up to 21,000 psi (144,795 kPa) through the nozzle assembly. Fuel is supplied at approximately 60 psi (414 kPa) through fuel rails in the cylinder heads. Return fuel is plumbed from fittings at the front of each cylinder head to a regulator block, which contains a piston/spring-type regulator valve that maintains pressure to approximately 40 psi (276 kPa). A de-aeration bleed orifice between the fuel filter and the regulator block vents air trapped in the fuel filter. Most of the fuel from the regulator is re-circulated to the inlet of the piston (high-pressure) stage of the transfer pump. Fuel return to the tank is limited by the fuel filter bleed orifice and an 0.0008 mm (0.020 in.) fuel return bleed orifice. This prevents the fuel from overheating in the tank. This system is controlled by PCM and is composed of the unit injectors, the high pressure lubricating oil system, the injection driven module, and the fuel supply system.

Task F2.7 Perform on-engine inspections, tests, and adjustments on unit injector electronic controls.

Injector Response Time and Pulse Width

Injector response time (IRT) and pulse width (PW) are explained earlier in this chapter. Both are indicators of the overall electrical integrity and power balance of the engine. Both parameters can be displayed on any EST. It should be noted that when attempting to correlate PW, which is measured in milliseconds, with engine crank degrees, that as an engine is accelerated from idle speed to high idle speed, the number of crank angle degrees the engine passes through in one millisecond will triple. So do not be surprised if the PW average decreases as an engine is accelerated.

Injector Cutout Test

There is no appropriate way to mechanically short out a DDEC EUI, so when a cylinder misfire is being diagnosed, the ECM electronics perform this task by electronically cutting out the EUI in sequence and analyzing the performance effect. When a ECM commands an engine to run a specific rpm, and an injector cutout test is performed, first the average duty cycle of the EUI in milliseconds is displayed: if the engine rpm is to be maintained during the test (this test is normally run at idle or 1,000 rpm), as each EUI is cutout electronically by the ECM, the PW of the remainder of the injectors will have to be lengthened if engine rpm is held at the test value. This would be true until a defective EUI was cutout: in this case, there would be no increase in the average PW of the operating EUI. If the example of a Series 60 engine were used with a dead EUI the following would occur if an injector cutout test sequence were set with the engine running at 1,000 rpm. To run the engine at 1,000 rpm, the 5 functioning EUI would produce fueling values read in PW milliseconds which would be averaged on the display. When one of the 5 functioning EUI is electronically cutout by the ECM, only 4 EUI would be available to run the engine at the test speed of 1,000 rpm, causing the average PW to increase. As each functioning EUI was cut out in sequence, the average PW would have to increase to maintain the test rpm. However, when the defective EUI was cut out, there would be no change in the average PW because it was based on the engine running on 5 cylinders.

Task F2.8 Inspect, adjust, repair/replace electronic air/fuel ratio controls, circuits, and sensors.

Cylinder misfire—check for a problem that can be read electronically first on electronically managed engines. Perform an electronic cylinder cutout test using a PC or ProLink if that option is in the software diagnostics. With non-electronic engines, short out injectors in sequence using a rocker claw/lever for MUI and PT injectors, and by cracking (loosening the line nut) high pressure pipes in port-helix metering pumps.

Task F2.9 Inspect, test, adjust, repair/replace engine governor electronic controls, circuits, and sensors.

When the port-helix metering injection pump is managed by a computer, engine governing depends on how the ECM is programmed. The governor housing attached to the rear of the injection pump is replaced by a rack actuator housing consisting of switched ECM output devices and sensors. Primarily, the housing contains a rack actuator, either a linear magnet (proportional solenoid) controlled directly by the ECM or an electric-over hydraulic device (engine oil acts as the hydraulic medium), also ECM controlled. The rack actuator sets rack position on ECM command and nothing more. A rack position sensor reports the exact rack position to the ECM.

Additionally, the rack actuator housing may house other sensors to report rotational speed (pulse wheel fitted to rear of camshaft) engine position and timing data to the ECM. Rack actuator housings are not governors. Where a rack actuator housing is fitted to an in-line, port-helix metering injection pump, the engine governing functions are undertaken by the software programmed into the fuel management ECM.

Electronic governing offers infinitely more control over engine fueling and it allows OEMs to meet statutory noxious emissions requirements and achieve better fuel economy, but it additionally can be easily programmed/reprogrammed with customer data to tailor the engine for varying engine and chassis applications. This subject matter will be dealt with in detail elsewhere in this text in sections relating to partial authority, electronic engine management.

Task F2.10 Inspect, test, adjust, repair/replace engine electronic fuel shut-down devices, circuits, and sensors.

Engine Overspeed

At engine derate, the operator is alerted by the flashing light and engine power and speed deration will occur incrementally depending on the severity of the condition. The engine will not actually shutdown unless the ECM is programmed with the engine shutdown option. The engine protection shutdown option will automatically shutdown the engine when a fluid pressure or temperature out of normal range problem occurs that could lead to catastrophic damage. Should this occur, the dash engine protection lamp will flash for 30 seconds prior to the shutdown. Key off is required to restart the engine but if the shutdown trigger persists, the engine will be shutdown again after 30 seconds.

Task F2.11 Inspect, test, adjust, repair/replace electronic safety shut-down devices, circuits, and sensors.

The ECM/PCM is programmed with engine management logic to fuel the engine through variable speeds and loads, while at the same time monitoring both engine and chassis operating conditions and continually performing self diagnosis. DDEC III features EEPROM (electronically erasable programmable read only memory) which qualifies the master engine management program in ROM (read only memory) with application specific detailing such as engine governing, torque profile shaping, cold start logic, and engine protection strategy. The results of ECM arithmetic, logic, and fetch and carry computations in the processing cycle are converted to actuator or control signals that switch the EUI. Command pulses can be referred to as duty cycle or PW (pulse width) and are measured in milliseconds. Duty cycle or PW is the time within the EUI cam actuated downstroke in which fuel is actually being pumped. PW is calculated and controlled by the ECM; it is effected by the injector driver unit of the ECM known as the EDU, which energizes the control solenoids of the EUI.

Engine Protection

DDEC III monitors a range of vehicle and chassis functions and is programmed with engine protection strategy that may derate or shutdown the engine when a running condition could result in catastrophic failure. The operator is alerted to system problems by illuminating dash lights, the check engine light (CEL), and the stop engine light (SEL). The CEL is illuminated when a system fault is logged to alert the operator that DDEC III fault codes have been generated. The SEL is illuminated when the ECM detects a problem that could result in a more serious failure. DDEC III may also be programmed to shut down or ramp down to idle speed after such a problem has been logged. A shut down override switch known as an SEO or stop engine override switch will permit a temporary override of the shut down command.

Engine Protection Strategy

Engine protection strategy can be programmed to 3 levels of protection beginning at a low level of protection, essentially a driver alert, and culminating with a complete shutdown of the engine. The level of protection is selected by the vehicle owner and programmed to the DDEC ECM as a customer option. As a customer option, the failure strategy can be changed at any time by a technician equipped with the access password and an EST with the DDEC software.

Parameters monitored:

Low-coolant pressure
High-coolant temperature
Low-oil pressure
Low-coolant level
High-oil temperature
Auxiliary digital input

When DDEC III detects a fault that may result in catastrophic engine damage, it can be programmed to react to the problem in three ways:

1. Warning only. The CEL and SEL will illuminate but it is left up to the operator to take action to avoid potential engine damage. No power or speed reduction will occur. Engine protection is therefore left entirely to the discretion of the operator.

2. Ramp down. The CEL and SEL will illuminate when a fault is detected and the ECM will reduce power (fuel quantity injected) over a thirty-second period after the SEL illuminates. The maximum power value (fuel quantity) maintained for this initial reduction is that operative just prior to the logging of the fault condition. A diagnostic request/stop engine override switch is required when this engine protection option is selected.

3. Shutdown. Both the CEL and SEL will illuminate and the engine will shutdown 30 seconds afterwards. Once again, a diagnostic request/stop engine override switch is required if this option is selected.

Stop Engine Override Options

Three types of stop engine override are available:

1. Momentary override—resets the 30-second shutdown timer, restoring engine power to the value operative at the moment the SEL was illuminated. To obtain subsequent overrides the switch must be recycled after 5 seconds. DDEC III will record the number of overrides actuated after the triggering fault occurrence.

2. Continuous override—option 1. This stop engine override option is used when a vehicle needs full power during a shutdown sequence. Full power capability is maintained for as long as the override switch is pressed. It is intended for coach applications only.

3. Continuous override—option 2. This stop engine override option is used primarily for industrial and construction applications. It essentially disables the engine protection system until the ignition key is cycled.

CELECT Plus Engine Protection System

The CELECT Plus monitors critical engine sensors and will log fault codes when an out of normal condition is determined. If an out of range condition is detected in a critical sensor an engine derate action may be initiated. The operator will be alerted by flashing the appropriate dash light. The engine protection system monitors:

Coolant temperature
Oil temperature
Intake manifold temperature
Oil pressure
Coolant level

Task F2.12 Inspect and test power and ground circuits and connections; determine needed repairs.

With modern electronic modules and low-level signals, proper grounding between modules has become extremely important. Even a resistance of one to two ohms between ground can cause a fault to occur. Test grounds without power applied using a digital ohmmeter between the ground lead and a clean common ground. Do not use a standard ohmmeter as the internal battery could cause damage to the module. Test grounds with power applied using a digital voltmeter between the ground lead and a clean common ground. Do not use a standard voltmeter because its internal resistance can affect the internal circuit of the module. Once defective ground has been detected, clean the ground connection and treat it with conductive grease. Fasten all leads securely and connect a suitable wire between common grounds. Re-test all circuits to ensure proper operation.

Task F2.13 Inspect and replace electrical connector terminals, seals, and locks.

The ECM harness electrically connects the ECM to other modules in the vehicle engine and passenger compartments. Wire harnesses should be replaced with proper part number harnesses. When signal wires are spliced into a harness, use wire with high temperature insulation only. With the low current and voltage levels found in the system, it is important that the best possible bond at all wire splices be made by soldering the splices. Molded on connectors require complete replacement of the connector. This means splicing a new connector assembly into the harness. Use care when probing a connector or replacing terminals in them. It is possible to short between opposite terminals. If this happens to the wrong terminal pair, it is possible to damage certain components. Always use jumper wires between connectors for circuit checking. NEVER probe through the weather pack seals. A connector test adapter kit is used to probe terminals during diagnosis. A fuse remover and test tool should be used when removing a fuse and to adapt a fuse, hold with a DMM meter for diagnosis. When diagnosing, open circuits are often difficult to locate by sight because of oxidation, or the terminal alignment is hidden by the connectors. Merely wiggling a connector on a sensor in the wiring harness may correct an open circuit condition. This should always be considered when an open circuit or failed sensor is indicated. Intermittent problems may also be caused by oxidized or loose connections. Before making a connector repair be certain of the type of connector. Weather pack and pull to seal connectors look similar but are serviced differently. Some connectors, such as the coolant sensor, use terminals called micro connectors. They are changed by inserting a fine pick tool through the front of the connector to bend a locking tab on the terminal. To remove, bend the tab toward the terminal, pull the terminal and wire out of the connector. To install, insert the tool into a new connector and bend the tab away from the terminal. Molded plugs and connectors are factory fabricated and must be spliced into the harness. Ensure correct splice procedures are used to make a good electrical connection. "Pull to seat" terminals. To install a terminal on a wire, the wire is first inserted through the seal and connectors, then the terminal is crimped on the wire and the terminal is pulled back into the connector to seat it in place. To remove this type of terminal:

- Slide the seal back on the wire.
- Use an insertion tool to release the terminal-locking tab.
- Push the wire and terminal out through the connector. If re-using the terminal, reshape the locking tang.

A weather pack connector can be identified by a rubber seal at the rear of the connector. This connector is in the engine compartment and protects against moisture and dirt that could create oxidation and deposits on the terminals. This protection is important because of the very low voltage and current levels found in the electronic system. To repair a weather pack terminal, use a special tool to remove the pin and sleeve terminals. If removal is attempted with an ordinary pick there is a good chance that the terminal will be bent or deformed. Unlike standard blade type terminals these terminals cannot be straightened once they are bent. Make certain that the connectors are properly seated and all of the sealing rings are in place when connecting leads. The hinge-type flap provides a back-up or secondary locking feature for the connector. They are used to improve the connector reliability by retaining the terminals if the small terminal lock tangs are not positioned properly. Weather pack connections cannot be replaced with a standard connection. Instructions are provided with weather pack connector and terminal packages.

Task F2.14 Connect computer programming equipment to vehicle/engine; access and change customer parameters; determine needed repairs.

Modern automotive electronic systems have stored information pertaining to customer options. These are used to change the operational characteristics of the control modules. To access and re-program this information, connect a scan tool to the diagnostic connector usually located under the dash on the driver's side. This will provide a link with all programmable modules either directly or over a network protocol line. These links must be automatically compatible to generic scan tools by standards established by all automotive manufacturers and international governments. Each manufacturer will provide necessary codes to make required changes to achieve proper operation of the vehicle and customer loaded requirements. A technician using a generic scan tool must use manufacturer's manual listed codes to make changes of the manufacturer's approved scan tool.

The powertrain control module (PCM) reset allows the scan tool to command the PCM to clear all emission-related diagnostic information. When resetting the PCM, a code will be stored in the PCM until all the OBD II system monitors or components have been tested to satisfy a drive cycle. Without any other faults occurring. The following events occur when PCM reset is performed:

- Clears the number of diagnostic trouble codes (DTCs).
- Clears the DTCs.
- Clears the scan tool freeze frame data.
- Resets status of the OBD II system monitors.
- Sets the code.

Resetting keep alive random access memory (RAM):

Disconnect the battery ground cable for a minimum of five minutes. Resetting keep alive RAM will clear learned values the PCM has stored for adaptive systems such as idle and fuel trim. After keep alive RAM has been reset, the vehicle may exhibit certain driveability concerns. It will be necessary to drive the vehicle to allow the PCM to relearn values for optimum driveability and performance.

Continuous memory diagnostic trouble code (DTC) access:

Continuous memory DTC access is also a self-test of the powertrain control system. Unlike the Key On Engine Off (KOEO) and Key On Engine Running (KOER) self-tests, which can only be activated on demand, the continuous self-test is always active. The test consists of all the On Board Diagnostics (OBD II) monitors and the comprehensive component monitor, and is designed to detect failures contributing to driveability or emission concerns. When using a scan tool it is possible to select "Retrieve Continuous Codes" or the malfunction indicator lamp (MIL) DTCs which activate the "check engine" or "service engine soon" indicator in the instrument clusters and are only generated for faults with the emission controls. Generally, non-MIL DTCs is for controls other than the emission system (i.e., A/C cycling switch or TR sensor faults). Any scan tool that meets OBD II requirements can access continuous memory to retrieve MIL DTCs. However, when performing KOEO or KOER self-test or retrieving all continuous memory DTCs (MIL and non-MIL) with a generic scan tool, it may require additional manually entered commands. Most vehicle manufacturers provide access codes which are used for entering commands. When using a generic scan tool, it is also possible to access MIL DTCs that are pending and have not actually turned on the MIL. A pending DTC applies to a fault set during the first drive after a PCM reset that has not turned on the "malfunction indicator lamp." Pending DTC retrieval is useful for a repair verification without the need to drive the vehicle through more than one drive cycle. Previously, a MIL DTC was not accessible by a generic scan tool until the MIL illuminated.

G. Starting System Diagnosis and Repair (4 Questions)

Task G1 Perform battery state-of-charge test; determine needed service.

Disconnect the battery cables from the terminals. Install a voltmeter and battery load tester to the adapters. Remove the surface charge from recently charged batteries by applying a 300-ampere load across the ports for 15 seconds. Do not remove the surface charge from batteries, which have been in storage. Turn the load off and wait 15 seconds for the battery to recover. Apply the specified load selected from the chart in ASE specifications. Observe the battery voltage after 15 seconds with the load connected, then turn off the load. If the battery voltage does not drop below the minimum voltage as shown in the following voltage and temperature chart, the battery is good and should be returned to service. The battery temperature must be estimated by feel and by the temperature the battery has been exposed to for the preceding few hours. If the battery voltage drops below the minimum voltage listed, replace the battery.

Voltage and Temperature Chart

TEMPERATURE	MINIMUM VOLTAGE
21°C (70°F & Above)	9.6
10°C (50°F)	9.4
–1°C (30°F)	9.1
–10°C (15°F)	8.8
–18°C (0°F)	8.5

Task G2 Perform battery capacity (load, high-rate discharge) test; determine needed service.

The battery discharge rate for a capacity test is usually one-half of the cold cranking rating. The battery is discharged at the proper rate for 15 seconds to eliminate the surface charge. Then, after about 1–2 minutes, the test is repeated. The battery voltage must remain above 9.6V with the battery temperature at 70°F or above.

To perform a battery drain test, clip one end of a jumper wire to the negative battery terminal and other end of the jumper wire to a digital multimeter or equivalent. Clip a second jumper wire to the end of the negative battery cable and to the multimeter.

- Set the multimeter on the DC, MA and take the reading with the engine control switch and all accessories off.
- If current draw is too high, check the system for causes, such as a shorted wire or a compartment lamp that does not shut off when it should.

Task G3 Charge battery using slow or fast charge method as appropriate.

One uses an open circuit voltage test to determine a battery's state of charge. Follow these steps to test voltage. If the battery has just been charged or has recently been in for service, the surface charge must be removed before an accurate voltage measurement can be made. To remove the surface charge, crank the engine for 15 seconds. Do not allow the engine to start. If the battery is out of the vehicle, a load test of half the rated cold cranking amps for 15 seconds will remove the surface charge. After loading the battery by either method, let the battery rest for 15 minutes. Connect a voltmeter across the battery terminals and observe the reading. If the reading is 12.6 volts or more, the battery is fully charged. If the voltage reading is below 12.6 volts, the battery needs to be charged.

The built-in hydrometer provides a guide for battery testing and charging. In testing, the green dot means the battery is charged enough for testing. The hydrometer reading is not an indication of a good, bad, or fully-charged battery. If the colored dot is not visible and the hydrometer has a dark appearance, it means the battery must be charged before the test procedure is performed. Green is an indication of a 65 percent charge. Dark is an indication of less than a 65 percent charge. Light yellow or clear indicates a low electrolyte level. If the colored dot is clear or light yellow, this means that the fluid level is below the bottom of the hydrometer. This may be caused by prolonged charging, a broken case, tipping, or normal battery wearout. First, tap the battery and if the clear or light yellow remains, replace the battery.

If a side terminal battery or threaded top terminal battery is in the vehicle, connect the charger leads to the studs or nuts at the battery terminals. If a side terminal or threaded top terminal battery is disconnected from the vehicle electrical system for any reason, install appropriate terminal adapters. The following basic rules apply to any sealed battery-charging situation:

- Do not charge a battery if built-in hydrometer is clear or light yellow. Replace the battery.
- Charge rates between 3 and 50 amperes are satisfactory as long as spewing of the electrolyte does not occur or the battery does not feel over 125°F (52°C). If spewing occurs or the temperature exceeds 125°F (52°C), the charging rate must be reduced or temporarily halted to permit cooling. Estimate battery temperature by touching or feeling the battery case.
- The battery is sufficiently charged when the green dot in the built-in hydrometer is visible. No further charging is required. Tap the hydrometer lightly on top at hourly intervals during charging and see if the green dot appears.

Battery charging consists of a charge current in amperes for a period in hours. Thus, a 25-ampere charging rate for 2 hours would be a 50-ampere hour charge to the battery. In most cases, batteries whose load test values are less than 200 amperes will have the green dot visible after at least a 50-ampere hour charge. Most batteries whose load test values are greater than 200 amperes will have the green dot visible after at least a 75-ampere hour charge. In the event that the green dot does not appear after this amount of charging, continue charging for another 50- or 75-ampere hours. If the green dot still does not appear, replace the battery.

Task G4 Start a vehicle using jumper cables, a booster battery, or auxiliary power supply.

If the vehicle will not start due to a discharged battery, it can often be started by using energy from another battery—a procedure called "jump starting." For example, the vehicle has a 12-volt starting system and a negative ground electrical system. Make sure that the other vehicle also has a 12-volt starting system and that it is the negative (-) terminal which is grounded or attached to the engine block or frame rail. Its operator's manual may give you that information. Do not try to jump start if you are unsure of the other vehicle's voltage or ground, or if the other vehicle's voltage and ground are different from your vehicle. Diesel engine vehicles have more than one battery because of the higher torque required to start a diesel engine. This procedure can be used to start a single battery vehicle from any of the batteries of the diesel vehicle. However, at low temperatures it may not be possible to start a diesel engine from a single battery in another vehicle.

Position the vehicle with the good (charged) battery so that the booster (jumper) cables will reach, but never let the vehicles touch. Also, be sure the booster cables do not have loose or missing insulation.

In both vehicles:

- Turn off the ignition (engine control switch) and all lamps and accessories except the hazard flasher or any lamps needed for the work area.
- Apply the parking brake firmly. Shift the automatic transmission or manual transmission to Neutral.
- Making sure the cable clamps do not touch any other metal parts, clamp one end of the first booster cable to the positive (+) terminal on one battery, and the other end to the positive terminal on the other battery. Never connect (+) to (-).

- Clamp one end of the second cable to the negative (-) terminal of the good (charged) battery. Make the final connection to the frame rail, chassis, or to any solid, stationary, unpainted metallic object on the engine at least 18 inches (450 mm) from the discharged battery or some other well-grounded point if the battery is mounted outside the engine compartment. Make sure the cables are not on or near pulleys, fans, or other parts that will move when the engine is started.
- Start the engine of the vehicle with the good (charged) battery and run the engine at a moderate speed for several minutes. Then start the engine of the vehicle that has the discharged battery.
- Remove the booster cables by reversing the above installation sequence exactly. While removing each clamp, take care it does not touch any another metal while the other end remains attached.

Task G5 Inspect, clean, repair/replace battery, battery cables, and clamps.

If battery voltage is disconnected from a computer, the adaptive memory in the computer is erased. In a powertrain control module (PCM), disconnecting power may cause erratic engine operation or erratic transmission shifting when the engine is restarted. After the vehicle is driven for about 20 miles, the computer relearns the system, and normal operation is restored. If the vehicle is equipped with personalized items such as memory seats or mirrors, the memory will be erased in the electronic module that controls these items. Radio station presets will also be erased. A 12V power supply from a dry cell battery can be connected through the cigarette lighter or power point connector to maintain voltage to the electrical system when the battery is disconnected.

A battery can be cleaned with a baking soda and water solution. Always wear hand and eye protection when servicing batteries and their components. If the built-in hydrometer indicates light yellow or clear, the electrolyte level is low, and the battery should be replaced. The low electrolyte level may be caused by a high voltage regulator setting that causes overcharging. When disconnecting battery cables, always disconnect the negative battery cable first.

It is always wise to spray the cable clamps with a protective coating to prevent corrosion. A little grease or petroleum jelly will also prevent corrosion. Also available are protective pads that go under the clamp and around the terminal to prevent corrosion.

When replacing a battery, it is very important to check the charging system. A faulty alternator or regulator could cause a discharged or overcharged condition, resulting in a damaged battery. The cables should be inspected and the terminals cleaned and replaced only as necessary.

When removing a battery from a vehicle, the battery ground cable is always removed first and the positive cable last. The order is reversed when installing the battery and the cables.

Task G6 Inspect, test, and reinstall/replace starter relays, safety switch(es), and solenoids.

Most starter relays are an integral part of the starter and are replaced as a unit. Other starter relays can attach either to the starter as a separate unit or to the fender well. Removal of these types of starter relays entails removing the two positive leads and the control wires. The starter relay is an electrical high current switch, and as such, it does not have a ground lead. Any time maintenance is to be performed on the starter, remove the ground (negative lead) from the battery to prevent accidental shorting of the wiring.

Task G7 Perform starter current draw test; determine needed repairs.

High starter current draw, low cranking speed, and low cranking voltage usually indicate a faulty starter motor. Low current draw, low cranking speed, and high cranking voltage usually indicate excessive resistance in the starting circuit. The starting system circuit carries the high current flow within the system and supplies power for the actual engine cranking. Starting circuit components are the battery, battery cables, magnetic switch or solenoid, and the starter motor.

The cranking current test measures the amount of current, in amperes, that the starter circuit draws to crank the engine. This amperage reading is useful in isolating the source of certain types of starter problems.

An inoperative starter, along with a high current draw, is always an indication of a direct short to ground. Any opens, failed switches, springs, or relays would cause an operative failure and a no-current condition.

Task G8 Perform starter circuit voltage drop tests; determine needed repairs.

To check the resistance in the ground circuit, use a voltmeter to check voltage drops. Measure the voltage drop across each component in the starter circuit to check the resistance in that part of the circuit. Read this voltage drop while the starter motor is operating. For example, connect the voltmeter leads between the positive battery terminal and the positive cable on the starter solenoid; crank the engine to measure the voltage drop across the positive battery cable. A ground circuit voltage drop test is performed by connecting a voltmeter between the ground side of the battery and the starter housing or base. Then, the engine is cranked and the voltage read. Voltage should not exceed 100 millivolts.

Excessive resistance caused by poor terminal connections and partial short circuits through worn cable insulation will result in an abnormal voltage drop in the starter cable. Low voltage at the starter will prevent normal starter operation and cause hard starting.

- Check the voltage drop between ground (negative battery terminal) and the vehicle frame. Place one prod of the test voltmeter on the grounded battery post (not on the cable clamp) and the other on the frame. Operate the starter and note the voltage reading.
- Check the voltage drop between the positive battery terminal and starter terminal stud with the starter operating.
- Check the voltage drop between the starter housing and frame with the starter operating.
- If the voltage drop in any of the above is more than 1.0 volt, there is excessive resistance in the circuit. To eliminate resistance, the cables should be disconnected and connections cleaned. If cables are frayed or the clamps corroded, the cables should be replaced. When selecting new cables, be sure they are at least as large as the ones being replaced.

Task G9 Remove and replace starter.

Disconnect the negative cable of the battery to prevent shorting while performing maintenance. Disconnect the positive lead and ground cable from the starter motor. Remove the leaf shield, if installed. Remove control wires from the solenoid, if attached. Remove starter mounting bolts and starter motor. Installation is similar, except when installing the motor in the flywheel housing it may need align shims. Always make connection of the ground cable to the battery the last step.

H. Engine Brakes (2 Questions)

Task H1 Inspect, test, and adjust engine brakes.

When the compression brake is energized, it causes the engine to perform like a power-absorbing air compressor. The brake opens the exhaust valves before the compression stroke are complete and combustion does not occur in the cylinder. The compressed air is released into the engine exhaust system and energy is not returned to the engine through the power stroke. The following describes the process of opening the exhaust valves to release the compressed air from the cylinder.

When the solenoid valve is energized, engine-lubricating oil flows under pressure through the control valve. Then the oil flows to both master piston and slave piston. The oil pressure causes the master piston to move down against the adjusting screw of the injector rocker lever.

The push rod moves the adjusting screw end of the rocker lever up during the injection cycle. This causes the adjusting screw to push against the master piston. The movement of the master piston causes high oil pressure in the oil passage from the master piston to the slave piston. The ball check valve in the control valve holds the high pressure in the oil flow from the master piston to the slave piston.

The high pressure in the oil flow causes the slave piston to move down against the crosshead and opens the exhaust valves. The exhaust valves open as the piston moves to near the end of the compression stroke. The compression braking cycle is completed as the compressed air is released from the cylinder.

Task H2 Inspect, test, adjust, repair/replace engine brake control circuits, switches, and solenoids.

Engine brakes are designed to complement the main vehicle brake system, not replace it. An engine compression brake's ability to absorb power is low compared to the brake capacity of the vehicle, however, a large percentage of vehicle braking uses less than one fifth of the total vehicle capacity so engine brakes can greatly extend the life of the vehicle foundation brakes.

Most engine brakes are controlled by an electric circuit that requires that a series of switches are closed: at minimum the control switch, a clutch switch, and an accelerator or governor switch must be closed. Internal compression brakes use the electric control circuit to manage the brake's hydraulic circuit.

The operating principle of an internal engine compression brake is to make the piston perform the work of compressing the cylinder air charge and then negate or cancel the power stroke by releasing the cylinder charge by opening the exhaust valves at TDC on the compression stroke. This changes the engine's role from that of a power producing pump to that of a power absorbing pump.

The mechanical force used to time and actuate the hydraulic circuit of a Jacobs type internal engine compression brake is rocker movement: this may be the movement of the PT, MUI, or EUI rocker, or in cases where hydraulic injectors are used, movement of an exhaust valve rocker over a different cylinder.

The engine exhaust brake operates by choking down the exhaust flow: the retarding stroke of the piston is therefore the exhaust stroke. Exhaust brakes are managed electrically and actuated pneumatically.

Some engines use an internal engine compression brake in conjunction with an exhaust brake enabling both the upward strokes of the piston to be retarding strokes. This increases the engine braking capacity by a significant amount.

The Brake Saver uses the principle of a torque converter in reverse to absorb energy and retard vehicle speed. The Brake Saver uses engine oil as its retarding medium: the engine oil is provided to the hydraulic brake from a larger capacity oil sump by a dedicated section of the oil pump. All engine retarders operate at optimum efficiency when engine rotational speeds are highest.

Task H3 Inspect, repair/replace engine brake housing, valves, seals, screens, lines, and fittings.

The term engine brake can be used to describe retarder devices that use three distinct operating principles:

1. Internal engine compression brakes
2. External engine compression brakes
3. Hydraulic engine brakes

Internal Engine Compression Brakes

In the 4-stroke cycle diesel engine, the piston is required both to perform work and receive work: on its upward travel during the compression stroke, it compresses the cylinder air charge (performs work), following which it is forced through its downstroke by expanding cylinder gases (receives work). The internal engine compression brake operates by making the piston perform the work of compressing the air charge on the compression stroke and then negating (canceling) the power stroke by releasing the compressed cylinder gasses to the exhaust system (gas blowdown) somewhere around TDC. This has the effect of reversing the engine function, converting it from an energy producing pump to an energy absorbing compressor. All engine compression brakes use this principle of operation: the mechanisms used to actuate time and control the brakes vary. The braking powerflow of the vehicle begins at the drive axle wheels, extends through the transmissions and driveshafts, through the engine powertrain and the kinetic energy is converted to pressure: the potential energy of the compressed air is then dumped into the exhaust system. Braking efficiencies are highest when engine rpm is highest.

External Engine Compression Brakes

The operating principle of the external engine compression brake is not that different from the internal compression brake. External engine compression brakes are also known as exhaust brakes. They consist of a housing located downstream from the turbine housing discharge in the exhaust system: within the housing is a valve that when actuated chokes off the exhaust discharge. This restricts the exhaust gas flow and once again reverses the role of engine, converting it into an energy absorbing pump; however, in the internal engine compression brake the effective pumping stroke is the compression stroke, while in the exhaust brake, the effective retarding stroke becomes the exhaust stroke. Because of this, as each succeeding exhaust slug is unloaded into the exhaust manifold, the pressure rises and braking efficiency increases. Retarding efficiency is to some extent diminished by pressure loss to the intake during valve overlap. Operation of the exhaust brake is managed so that the engine is not fueled during braking and as with the internal engine compression brake, braking efficiencies are highest when the engine rpm in its higher range.

Hydraulic Engine Brakes

Many types of electric and hydraulic driveline brakes exist, but these are more appropriately dealt with when studying transmission and driveline components. However, the Caterpillar Brake Saver is a hydraulic retarder that is coupled to the rear of the engine and uses engine lubrication oil as its medium. A rotor is coupled directly to the engine crankshaft and so will rotate at any time the engine is running. The rotor is turned within the Brake Saver Housing which is coupled to the flywheel housing and to which the transmission is mounted. The engine flywheel is bolted through to the crankshaft, but because of the Brake Saver, the starter motor must crank the engine by means of a ring gear on a ring gear plate mounted to the crankshaft behind the brake saver rotor. The rotor is therefore driven by the crankshaft between the Brake Saver housing and a stator: when the Brake Saver is actuated, the housing is charged with pressurized engine oil. Vaned pockets on the rotor means that when the Brake Saver is charged with oil, the rotor encounters fluid resistance defined by the stator geometry and the rotor rotational speed. Once again, retarding efficiencies are greatest when rotational speed is greatest. The oil that acts as the Brake Saver hydraulic medium is engine oil, supplied from the engine sump by a section of the oil pump dedicated to changing the Brake Saver.

Engine Brake Operation and Control Circuits

Engine brakes typically use electric control switches which means that they may be actuated either directly by the driver or by the engine management ECM in "smart" cruise applications. When engine brakes are used in hydromechanically managed engines, the switching of the engine brake is usually electrical but the retarding effect must actuate either hydraulically or pneumatically. In such engines, a control circuit would require that a series of switches be closed before engine braking can be effected. The first of this series of switches would be the driver control switch which can be proportional depending on the system. The next would be located at the clutch: this must be necessarily fully engaged to operate engine braking. The final essential switch in the series (though any number of others may be installed) would be to ensure that the accelerator was not depressed, in other words, engine fueling is at zero (current engines) or at least at a minimum in older engines to limit the quantity of raw fuel dumped into the exhaust system. Current regulations require that no uncombusted fuel is discharged into the exhaust system so the rules of operation have changed in recent years. However, engine brakes used in many current engines are managed/monitored by the engine electronics. When an engine brake is electrically switched to the on position, the result depends on the type of engine brake used. Some of the common engine brakes are described in this next section:

Williams Exhaust Brakes

Williams exhaust brakes are classified as external engine compression brakes. Control of the brake is electric over pneumatic. The electrical circuit required to actuate the engine brake consists of three switches all of which must be closed: a control switch (dash mounted), a clutch switch (clutch must be fully engaged), and an accelerator switch (accelerator must be at zero travel). The sliding gate exhaust brake uses a pneumatically activated gate, actuated by chassis system pressure: the air supply to close the gate is controlled by an electrically switched pilot valve. An aperture in the sliding gate permits a minimal flow through the brake gate during engine braking. The butterfly valve version operates similarly.

Jacobs Compression Brakes

The Jacobs family of engine compression brakes is classified as internal compression brakes. People in the trucking industry use the term "Jake brake" to refer to any internal engine compression brake and indeed, this manufacturer dominates the market place designing their engine brakes for all the major OEMs. However, there are other manufacturers of similar devices and more evidence of this can be seen in imported truck engines. Jacobs also manufacture driveline retarders.

Sample Test for Practice

Sample Test

Please note the letter and number in parentheses following each question. They match the overview in section 4 that discusses the relevant subject matter. You may want to refer to the overview using this cross-referencing key to help with questions posing problems for you.

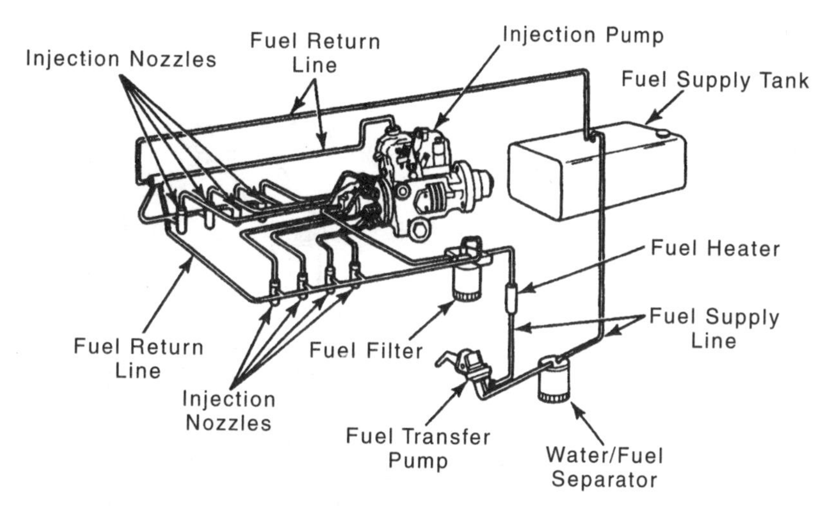

1. What type of fuel system is shown in the figure above?
   A. Distributor pump
   B. Jerk pump
   C. In-line pump
   D. Unit injector (F1.6)
2. A light film of slippery liquid covers the top of most engine compartment components, extending from the fan to the fire wall. Technician A says that the radiator may have a pinhole leak. Technician B says that the hydraulic fan motor hose may be leaking. Who is right?
   A. A only
   B. B only
   C. Both A and B
   D. Neither A nor B (A2)
3. Which of the following is LEAST likely to present a problem in an electrical/electronic circuit?
   A. Loose or improperly mated connector
   B. Wires with nicks or lumps in the insulation
   C. A melted or distorted electrical connector or wire
   D. Ground wires attached to an unpainted frame surface (A3)

4. A mechanical injector appears to be misfiring. Technician A says to use a metal bar like a stethoscope to find the bad cylinder. Technician B says to manually foul each injector and listen for the one that does not have an affect on engine performance. Who is right?
   A. A only
   B. B only
   C. Both A and B
   D. Neither A nor B (A4)
5. On startup, the exhaust appears as a white/light gray smoke for about 15 to 20 seconds, then turns almost clear. Technician A says that the engine may have a broken injector spring. Technician B says this is normal, but advises the operator to use an immersion heater when ambient temperature falls below 65°F (59°C). Who is right?
   A. A only
   B. B only
   C. Both A and B
   D. Neither A nor B (A12)

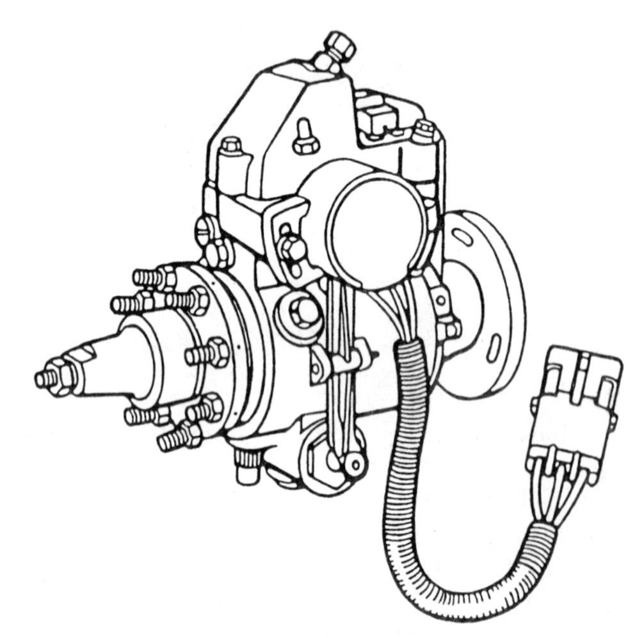

6. To adjust the fuel injection pump lever (FIPL) sensor shown in the figure above, all of the following must be done **EXCEPT:**
   A. connect a diagnostic tool to the vehicle.
   B. rotate the sensor after loosening screws.
   C. rotate the sensor until a steady tone can be heard from the test set.
   D. adjust the maximum throttle travel screw. (F2.2)
7. When using jumper cables to start a vehicle with a dead battery, always do which of these items?
   A. Connect (+) positive to (-) negative to create a series aiding circuit.
   B. Connect (+) positive to (+) positive and (-) negative to (-) negative to create a parallel circuit.
   C. Connect the (-) negative cable first and the (+) positive last.
   D. Make sure both vehicles are electronically grounded together before making any connections. (G4)

8. Technician A says to use an ohmmeter to check the resistance between (-) negative and ground while cranking the starter. Technician B says to check the voltage drop between (+) positive and the starter terminal stud while the starter is cranking. Who is right?
   A. A only
   B. B only
   C. Both A and B
   D. Neither A nor B (G8)
9. An engine cranks slowly and fails to start. Technician A says that the ambient engine temperature may be too low to obtain combustion temperature. Technician B says to check for a low battery or corrosion on the battery terminals and clean and charge as necessary. Who is right?
   A. A only
   B. B only
   C. Both A and B
   D. Neither A nor B (A11)
10. To determine battery capacity which specific measuring instruments are used?
   A. Multimeter
   B. Voltmeter and ammeter
   C. Ohmmeter and tachometer
   D. Ammeter and milli-ammeter (G2)

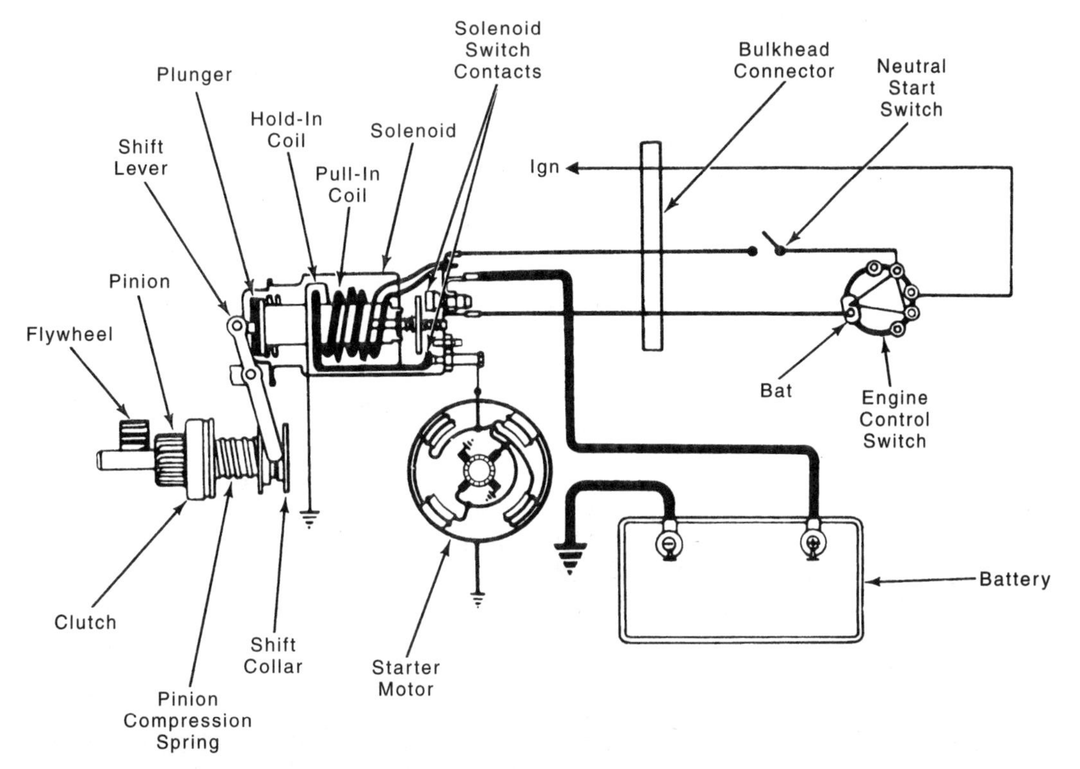

11. When removing the starter as shown in the figure above it is necessary to remove all of the following **EXCEPT:**
   A. starter ground cable.
   B. positive lead to starter/solenoid.
   C. battery ground cable.
   D. battery. (G9)

12. A driver complains that an engine had random loss of power while driving cross country. Further research reveals most of the driving was in mountainous areas. Technician A says the problem may have been caused by high altitude or low fuel level on a grade. Technician B says that the wind chill on the tank may have reduced the fuel temperature to less than 10°F (6°C) of the fuel's pour point. Who is right?
    A. A only
    B. B only
    C. Both A and B
    D. Neither A nor B (A12)
13. A vibration that has an occurrence rate equal to the engine speed is LEAST likely to be caused by:
    A. an out of balance turbocharger.
    B. broken or bent connecting rod.
    C. an out of balance or bent crankshaft.
    D. a cracked, chipped, or dented vibration damper. (A13)
14. Excessive surges in the coolant level could be caused by all of the following **EXCEPT:**
    A. a loose fan belt.
    B. a faulty fuel injector.
    C. a restricted radiator.
    D. a blown head gasket. (A14)
15. The LEAST likely cause for a low oil pressure indication is:
    A. a clogged oil pressure relief valve.
    B. a restricted oil pump suction tube.
    C. worn main bearings.
    D. a clogged oil filter. (A15)

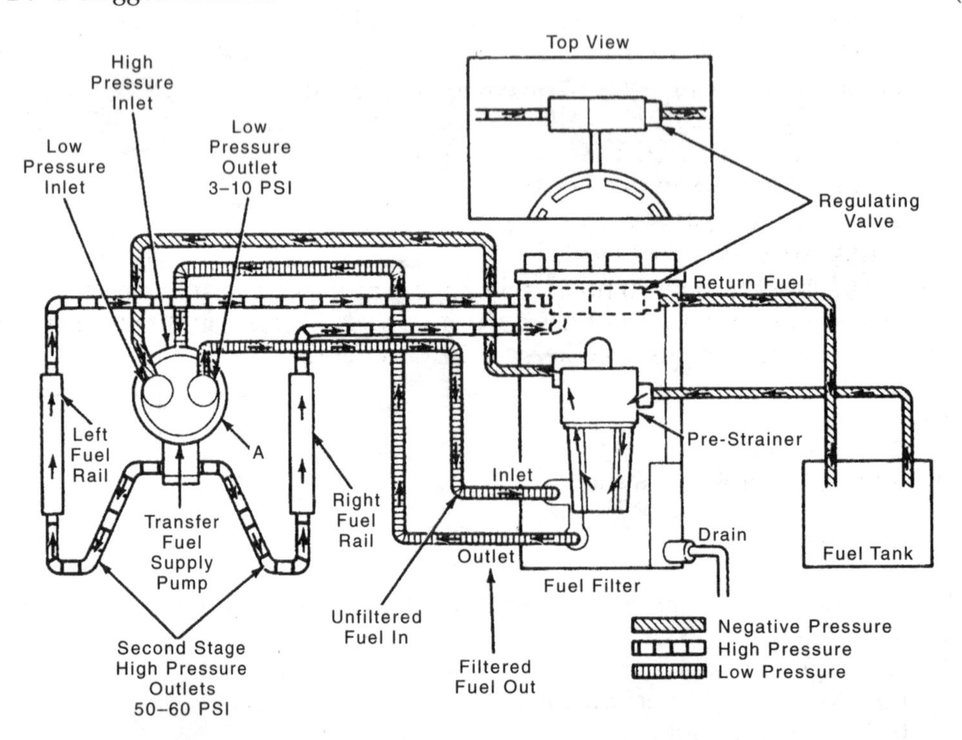

16. In the figure above of a unit injector system, the device labeled A is:
    A. a tandem fuel pump.
    B. a check valve.
    C. a pressure regulator.
    D. a pressure sensor. (F1.2)

17. To inspect an installed cylinder head for cracks, all of the following may be necessary **EXCEPT:**
    A. use a putty knife to remove excess gasket material from the head and inspect for visible cracks between cylinders and ports.
    B. check the torque of head bolts.
    C. use magnetic or dye crack detection to find small cracks that are not otherwise visible.
    D. remove excess carbon buildup by sanding. (B3)
18. A water pump mounting hole in the head has stripped threads. Technician A says that the head has to be replaced. Technician B says to install a helicoil. Who is right?
    A. A only
    B. B only
    C. Both A and B
    D. Neither A nor B (B2)
19. To check a cylinder warped block fire deck, Technician A says use a straightedge and feeler gauge for detection of a transverse or longitudinal warp. Technician B says to measure fire deck at 3, 6, 9, and 12 o'clock positions around each cylinder. Who is right?
    A. A only
    B. B only
    C. Both A and B
    D. Neither A nor B (C2)
20. When performing a cylinder head leakage test of fuel passages, which of the following is LEAST likely to be used?
    A. Discarded injectors
    B. Compressed air
    C. Threaded plug
    D. Injection pump (B11)
21. Which of the following is LEAST likely to be revealed during inspection and testing of valve springs?
    A. Springs that produce harmonic distortion
    B. Springs with uneven free height
    C. Cracks or brakes in springs
    D. Springs that are not square (B5)
22. An engine drops a valve 20 hours after an overhaul. Technician A says the probable cause was missing or defective valve seals. Technician B says the probable cause was improperly installed keepers. Who is right?
    A. A only
    B. B only
    C. Both A and B
    D. Neither A nor B (B6)

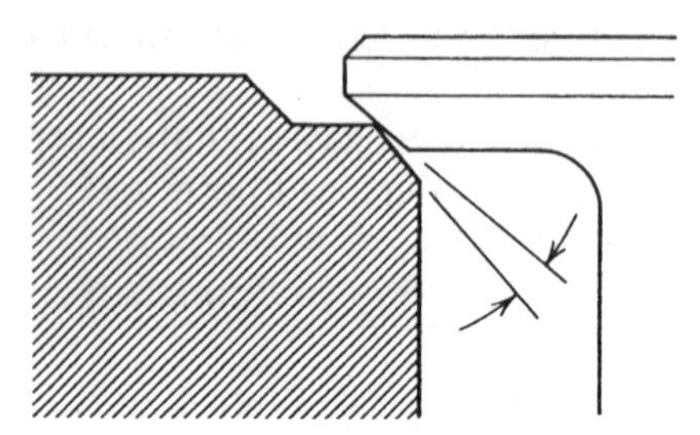

Intake Valve And Valve Seat

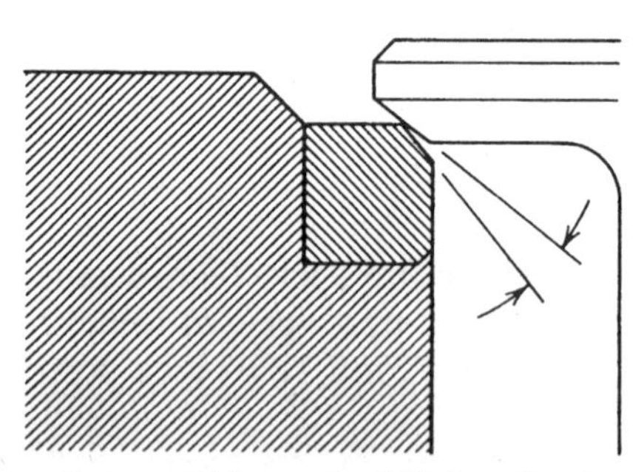

Exhaust Valve And Valve Seat

23. As shown in the figure above, Technician A says that the angle of the valve and seats must be ground the same if the valve has standard tappets. Technician B says that the seat angle must be 1 degree steeper when using a rotator. Who is right?
    A. A only
    B. B only
    C. Both A and B
    D. Neither A nor B (B9)
24. Technician A says that cupping of the valve head is normal and should not be used when considering a valve for replacement. Technician B says that valve margin measurements need not be taken for new valves. Who is right?
    A. A only
    B. B only
    C. Both A and B
    D. Neither A nor B (B8)
25. If an injector sleeve (tube) is found to be leaking, which of following procedures is LEAST likely to be required?
    A. Using the OEM process to remove and install the sleeve.
    B. Perfoming a cylinder head pressure test.
    C. Checking injector tip protrusion.
    D. Replacing the injector nozzle. (B11)
26. After placing pencil marks around the face of the valve about 1/8 inch (3 mm) apart and snapping the valve against its seat, all but two lines are found to be visibly broken. Technician A says the valve needs to be reground. Technician B says the valve and seat are within limits and will seal when spring pressure is applied. Who is right?
    A. A only
    B. B only
    C. Both A and B
    D. Neither A nor B (B9)
27. Which of the following tools is LEAST likely to reveal a defective cylinder sleeve?
    A. Inside micrometer
    B. Cylinder gauge
    C. Dial indicator
    D. Snap gauge or bore gauge (C5)

28. After wet-type cylinder sleeves have been removed, the packing ring area and counter bore lip show signs of light rust and scale. Technician A says this is normal and no additional action is required except ensuring antifreeze protection includes rust inhibitors. Technician B says to use 100 to 120 grit emery paper to remove rust and scale, and check for erosion and pitting. Who is right?
    A. A only
    B. B only
    C. Both A and B
    D. Neither A nor B (C2)
29. Cracks are present in the piston skirt. Technician A says to check the connecting rod and bottom of the cylinder line for damage caused by improperly installed connecting rods. Technician B says to check for insufficient piston-to-cylinder liner clearance. Who is right?
    A. A only
    B. B only
    C. Both A and B
    D. Neither A nor B (C14)

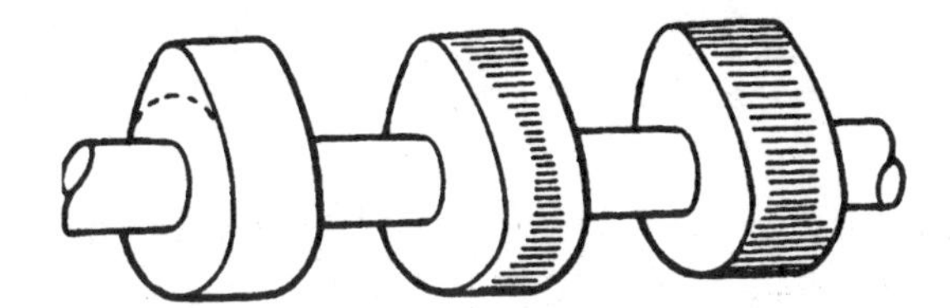

30. Two checks of the heel-to-toe measurement provide different readings, but they are still within specification, as shown in the figure above. Technician A says to check for excessive grooving of the cam lobe. Technician B says that as long as the readings are within specifications no further checks are required. Who is right?
    A. A only
    B. B only
    C. Both A and B
    D. Neither A nor B (C9)
31. Technician A says to check the bearings and crankshaft before cleaning the crank. Technician B says to check the crankshaft after cleaning. Who is right?
    A. A only
    B. B only
    C. Both A and B
    D. Neither A nor B (C10)
32. As a minimum, a variable speed governor must include all of the following **EXCEPT:**
    A. must be gear or belt driven from the engine.
    B. must control the injectors/fuel pump.
    C. must have throttle linkage control from the operator.
    D. must work independent of operator control. (F1.14)
33. Older mechanical temperature gauges consist of all of the following **EXCEPT:**
    A. fluid filled sensor.
    B. fluid filled tube.
    C. compensation resistor.
    D. hydraulic meter movement. (D7)
34. Of the following parts, which will LEAST likely be found on a pressure timed (PT) system?
    A. Gear pump
    B. Throttle linkage
    C. Pulsation damper
    D. Transfer pump (F1.8)

35. Technician A says the piston ring end gap should be checked after the rings are installed on the piston and the piston is in the cylinder. Technician B says the piston ring end gap is checked with only the rings in the cylinder. Who is right?
    A. A only
    B. B only
    C. Both A and B
    D. Neither A nor B (C16)

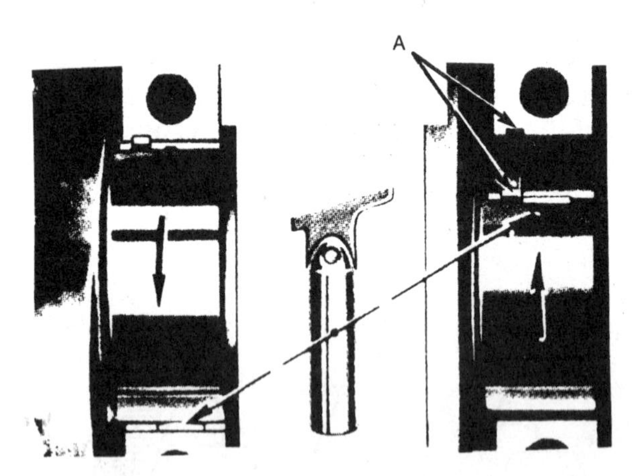

36. The notch in the bearing shell at point A in the figure shown above is used:
    A. to direct the oil evenly around the bearing.
    B. to indicate a defect in the manufacturing or handling of bearing shell.
    C. to indicate this is an oversized bearing shell.
    D. to keep the bearing shell from turning. (C12)
37. All of the following will cause uneven wear on the main bearings or stress fractures on the crankshaft **EXCEPT:**
    A. full load operation in excess of eight hours.
    B. ignition of fluid in the vibration damper.
    C. teeth missing from the ring gear.
    D. loose torque converter bolts. (C19)
38. A crankshaft is found to have fractures that may have been caused by improper alignment. Technician A says to use a master bar to check the main bore alignment. Technician B says to use an inside micrometer to check the main bore diameter for an out-of-round condition. Who is right?
    A. A only
    B. B only
    C. Both A and B
    D. Neither A nor B (C11)
39. An electrical oil pressure gauge does not indicate any pressure and the engine is operating well. Which of the following is at fault?
    A. The passage to the sensor is open.
    B. The wrong engine oil was installed in the engine.
    C. The wiring to the sensor is open.
    D. The wrong sensor was installed in this engine. (D1)
40. Technician A says regardless of the type of oil pump used, all mated surfaces must be checked for wear. Technician B says because oil pump gear receive so much wear, the oil pump should always be replaced during an overhaul. Who is right?
    A. A only
    B. B only
    C. Both A and B
    D. Neither A nor B (D2)

41. After inspecting an oil filter housing or its mounting, which of the following is LEAST likely to be a procedure?
    A. Visually check for cracks
    B. Visually check gasket surface for nicks
    C. Magna fluxing the housing and metal parts to detect small cracks
    D. Visually inspect housing passageways for obstructions and racks (D3)
42. Which of the following would indicate an oil cooler failure?
    A. Low oil pressure
    B. High oil pressure
    C. Milky gray sludge in oil filter
    D. Loss of engine oil (D4)

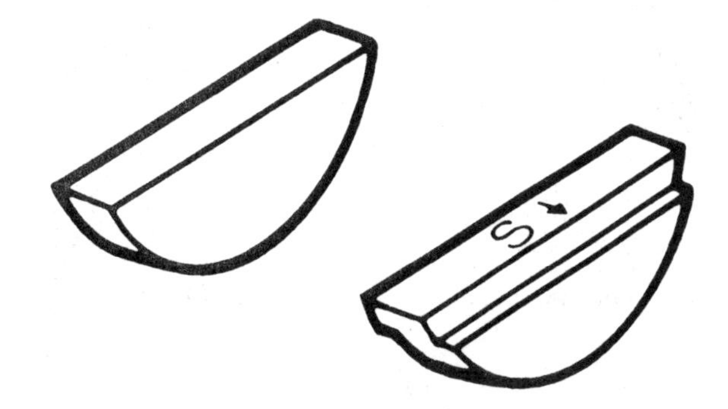

43. The figure shown above indicates the condition of a key on the right used on the camshaft gear. Technician A says the manufacturer used this to offset the timing and it must be replaced with the same type. Technician B says that the camshaft, gear, and key must be replaced because the camshaft became jammed and twisted and started to shear the key. Who is right?
    A. A only
    B. B only
    C. Both A and B
    D. Neither A nor B (C13)
44. Which of the following is LEAST likely to cause a cooling system failure?
    A. A water pump drive belt having at least two cracks within any 2 inch section of its material
    B. A drop of water from the pump bottom hole every five to ten minutes
    C. A lose clamp on a heater hose
    D. Rust or scum floating in the radiator (D11)
45. Technician A says back-flushing the radiator in the vehicle should be accomplished annually. Technician B says air locks usually occur when the cooling system is filled with the bleeder valves open. Who is right?
    A. A only
    B. B only
    C. Both A and B
    D. Neither A nor B (D9)
46. When checking the coolant condition with an SCA test strip, the technician finds that the coolant condition is higher than specification limits. Which of these items should the technician do?
    A. Add more antifreeze to increase the SCA.
    B. Continue to run the truck until the next PMI.
    C. Drain the entire coolant system and add the proper SCA mixture.
    D. Run the truck with no SCA additives until the next PMI. (D10)

47. In an injector-type fuel system, Technician A says the in-line pump pressurizes, meter's timing, and delivers fuel to injectors. Technician B says the injector lines between the pump and injection must be made of steel to handle high pressure fuel. Who is right?
    A. A only
    B. B only
    C. Both A and B
    D. Neither A nor B (F1.7)
48. A misfiring cylinder is diagnosed on a diesel engine with unit injectors. The technician replaces one injector. However, the engine continues to miss. Which of these could be the cause?
    A. A worn cam follower roller
    B. A restricted fuel filter
    C. High fuel pump pressure
    D. A leaking injector tube (A12)

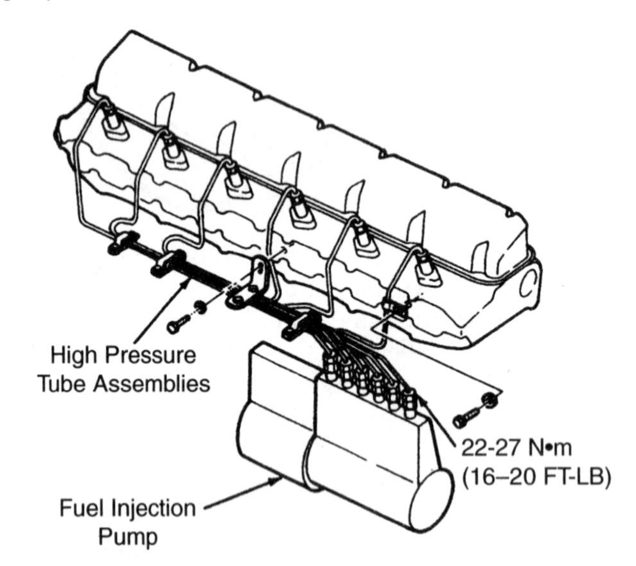

49. In the injector fuel lines in figure above, Technician A says they must be removed from the engine as an assembly. Technician B says remove only the line that needs to be replaced. Who is right?
    A. A only
    B. B only
    C. Both A and B
    D. Neither A nor B (F1.12)
50. An engine with a distributor-type injector pump cranks, but fails to start. What is the probable cause?
    A. A faulty injector
    B. A defective starter cutout relay
    C. A defective shutdown solenoid
    D. A leaky intake hose (F1.6)

51. The blades of a turbo intake are excessively worn. Technician A says this is an indication of a loose or damaged intake hose between the filter and the turbo. Technician B says that the aftercooler is not properly cooling the air before it enters the turbocharger, causing the compressor blades to become fatigued. Who is right?
    A. A only
    B. B only
    C. Both A and B
    D. Neither A nor B (E2)
52. Technician A says the air shutdown valve on a blower is used as an emergency shutdown device. Technician B says the air shutdown valve on a blower is used as a normal shutdown device. Who is right?
    A. A only
    B. B only
    C. Both A and B
    D. Neither A nor B (E9)
53. A medium duty diesel engine has high blowby gases and is leaking oil from several crankcase seals. Technician A says crankcase presure will increase if this condition is caused by an crankcase breather tube restriction. Technician B says you can use a pressure gauge graduated in inches of mercury to measure this pressure. Who is right?
    A. A only
    B. B only
    C. Both A and B
    D. Neither A nor B (E4)
54. A recently overhauled engine develops cylinder liquid lock. After clearing the liquid from the cylinder, inspecting the heads, and replacing the head gaskets the problem reoccurs. Technician A says the intercooler is defective. Technician B says the head has a hairline crack. Who is right?
    A. A only
    B. B only
    C. Both A and B
    D. Neither A nor B (E5)
55. A diesel engine develops too much boost at high engine rpm. Technician A says the oil bath air filter is dry. Technician B says the wastegate is malfunctioning. Who is right?
    A. A only
    B. B only
    C. Both A and B
    D. Neither A nor B (E1 and E2)

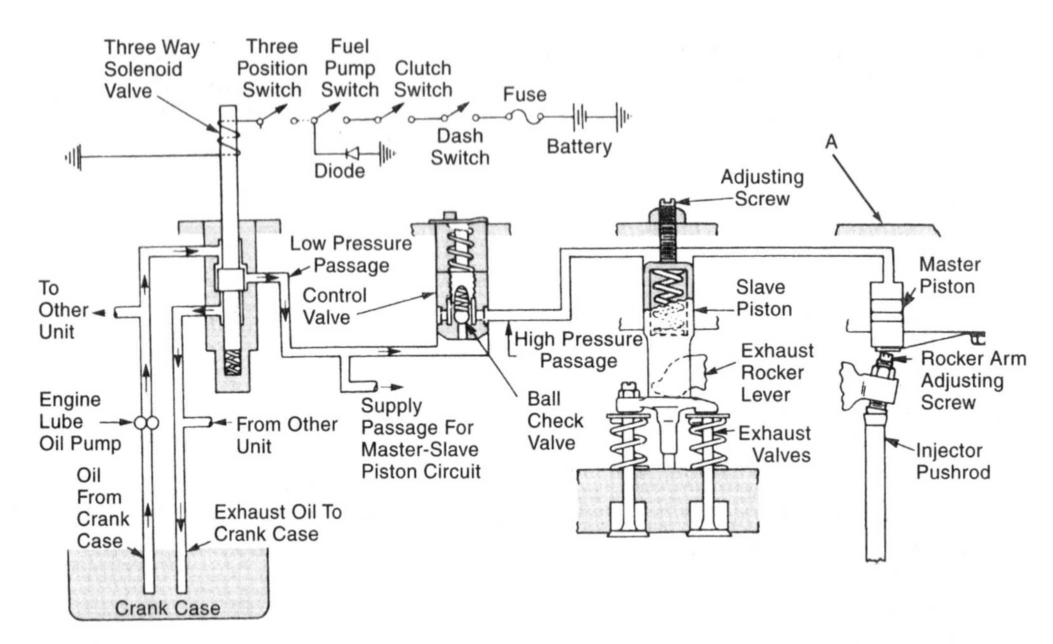

56. Technician A says that the device at A, in the figure above, will have seals or O-rings that must be inspected and replaced. Technician B says that for the exhaust brake to work properly, the adjustment at B must open exhaust valves before injection occurs. Who is right?
    A. A only
    B. B only
    C. Both A and B
    D. Neither A nor B (H1)
57. Technician A says that one of the switches at C in the figure above question 56, must be located at the fuel pump to detect when the accelerator is pressed. Technician B says that if the fuse in the circuit at C opens, the exhaust brake will engage. Who is right?
    A. A only
    B. B only
    C. Both A and B
    D. Neither A nor B (H2)
58. All of the following are steps in an air inlet restriction test **EXCEPT:**
    A. connect a manometer to the air intake.
    B. check the manometer with the air cleaner and duct removed.
    C. check the manometer reading with the air filter installed.
    D. place a piece of cardboard over the air intake to calibrate the manometer. (A10)
59. Technician A says that a gray or black smoke could be an indication of low blower air box pressure. Technician B says low blower air box pressure could be caused by leaky end plate gaskets. Who is right?
    A. A only
    B. B only
    C. Both A and B
    D. Neither A nor B (E3)

60. Technician A says to remove the surface charge from a recently charged battery, apply a 300-ampere load across the adapters for 15 seconds. Technician B says not to remove the surface charge from batteries which have been in storage. Who is right?
    A. A only
    B. B only
    C. Both A and B
    D. Neither A nor B (G1)
61. Excess exhaust back pressure may result in all of the following **EXCEPT:**
    A. lower engine power.
    B. higher exhaust temperature.
    C. higher intake vacuum.
    D. poor combustion. (A9)
62. Technician A says a loss of engine lubrication oil through the breather tube is an indication of excess crankcase pressure. Technician B says any crankcase pressure is a sign of trouble. Who is right?
    A. A only
    B. B only
    C. Both A and B
    D. Neither A nor B (A8)
63. In electronic diagnostics the acronym PID stands for:
    A. parameter identifier.
    B. part identification number.
    C. positive induction device.
    D. partial induction delivery. (F2.1)
64. Technician A says when removing low-pressure lines, inspect for damage to lines and connections. Technician B says to inspect seals and O-rings for serviceability and replace as necessary. Who is right?
    A. A only
    B. B only
    C. Both A and B
    D. Neither A nor B (F1.13)
65. When performing a snap pressure check and the pressure gauge reads 0 then:
    A. if the snap pressure is low it may be necessary to change the throttle restriction.
    B. reading will be the same for standard and AFC pumps.
    C. if the throttle travel is not correct, check to make sure the throttle is completely closed.
    D. if there is an incorrect low idle setting, change the governor. (F1.8)
66. Technician A says a scan tool is needed to clear a DTC from a vehicle's PCM. Technician B says DTCs can be cleared from the PCM by removing a battery cable. Who is right?
    A. A only
    B. B only
    C. Both A and B
    D. Neither A nor B (F2.1)
67. Technician A says most generic scan tools can access any OBD II vehicle MIL DTCs. Technician B says that vehicle manufacturers have to provide MIL DTC commands to be OBD II certified. Who is right?
    A. A only
    B. B only
    C. Both A and B
    D. Neither A nor B (A16)

68. When removing a starter relay, which of the following would LEAST likely be removed or disconnected?
    A. Positive lead to the relay
    B. Negative lead to the relay
    C. Relay
    D. Ground cable from battery (G6)
69. Technician A says not to charge a sealed battery if the hydrometer is clear or yellow. Technician B says the battery is charged when the built-in hydrometer dot is green. Who is right?
    A. A only
    B. B only
    C. Both A and B
    D. Neither A nor B (G3)
70. Technician A says that a radiator pressure cap with a 9 stamped on it should open to release pressure at between 8 psi (55 kPa) to 10 psi (69 kPa). Technician B says the same cap should open at about 0.6 psi (4.3 kPa) differential (vacuum) pressure. Who is right?
    A. A only
    B. B only
    C. Both A and B
    D. Neither A nor B (D12)
71. Technician A says when purging (bleeding) air from the fuel system it is necessary to pressurize the system with the hand primer. Technician B says purging (bleeding) air from the fuel system should be done any time fuel filters are changed. Who is right?
    A. A only
    B. B only
    C. Both A and B
    D. Neither A nor B (F1.4)
72. When using a voltmeter to perform a voltage drop test in a circuit, the leads should be connected in what way?
    A. To the battery terminals
    B. From the positive battery terminal to ground
    C. In series with the circuit being tested
    D. In parallel with the circuit being tested (G8)

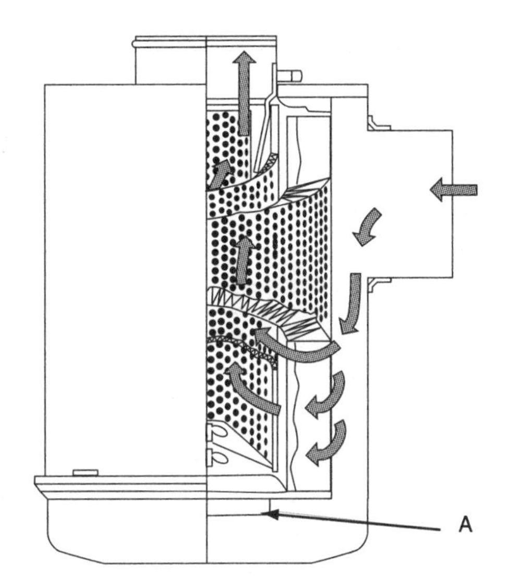

73. Technician A says the airflow shown in the figure above is an oil bath air filter system. Technician B says that the bottom canister (labeled A) is a dust collector and must be emptied when servicing the engine. Who is right?
    A. A only
    B. B only
    C. Both A and B
    D. Neither A nor B (E7)
74. Air shutters on a vehicle close automatically, but must be manually opened. What component of the system should you inspect?
    A. Shutter drive piston
    B. Shutter assembly springs
    C. Engine thermostat
    D. Shutter assembly thermostat (D14)
75. Technician A says to replace the fan if the blades are cracked. Technician B says bent fan blades can be straightened, even if badly bent. Who is right?
    A. A only
    B. B only
    C. Both A and B
    D. Neither A nor B (D13)
76. All of the following are used to control glow plugs in modern diesel engines **EXCEPT:**
    A. engine oil temperature.
    B. barometric pressure (BARO) sensors.
    C. powertrain control module (PCM).
    D. exhaust gas regulator (EGR). (E7)
77. With a capsule-type fluid starting system which of the following is LEAST likely?
    A. Capsules are not refillable.
    B. Check valves allow the plunger to develop fluid pressure for injection.
    C. Pump plunger is used to operate check valves.
    D. A solenoid valve is used to release starting fluid. (E8)
78. Technician A says that valve crossheads should be magna fluxed to check for cracks. Technician B says the valve crossheads should be checked for out-of-roundness and excessive diameter. Who is right?
    A. A only
    B. B only
    C. Both A and B
    D. Neither A nor B (B13)

79. An engine cranks, but fails to start and the operation of the primer pump fails to build pressure. Which of the following would cause this condition?
    A. Injector valve seized
    B. Fuel filter housing for leaks
    C. Restriction between tank and primer pump
    D. Lines between the prime pump and the injector pump for a clog (F1.3)

80. In the above figure, all of the following measuring devices can be used to check valve guides for proper clearance **EXCEPT:**
    A. a ball gauge.
    B. a snap gauge.
    C. a dial gauge.
    D. an outside micrometer. (B7)

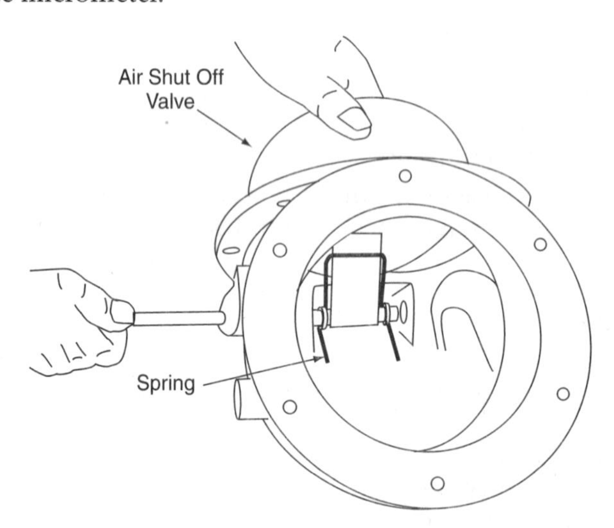

81. As shown in the figure above, Technician A says constant use of the emergency air shutdown valve is not harmful to the engine, but the fuel shutdown should be used instead. Technician B says emergency air shutdown valves should not be used on a regular basis. Who is right?
    A. A only
    B. B only
    C. Both A and B
    D. Neither A nor B (E9)

82. Technician A says to clean and blow out the grooves in the rocker cover with compressed air. Technician B says when using silicone for the valve cover gasket apply a light coat of the oil in the groove to make the gasket easier to install. Who is right?
    A. A only
    B. B only
    C. Both A and B
    D. Neither A nor B (C1 or B14)
83. Technician A says to adjust a fuel shutoff solenoid, the solenoid plunger must be bottomed out and checked to meet manufacturer's specifications. Technician B says to adjust the external stop screw to the point where it touches the rack control. Who is right?
    A. A only
    B. B only
    C. Both A and B
    D. Neither A nor B (F1.7)
84. Technician A says to properly test an engine block for cracks, it should be submerged in a tank of water heated to between 180–200°F (82–93°C). Technician B says an engine block can be checked for water leaks by pressurizing it for about two hours without loss of pressure. Who is right?
    A. A only
    B. B only
    C. Both A and B
    D. Neither A nor B (C3)

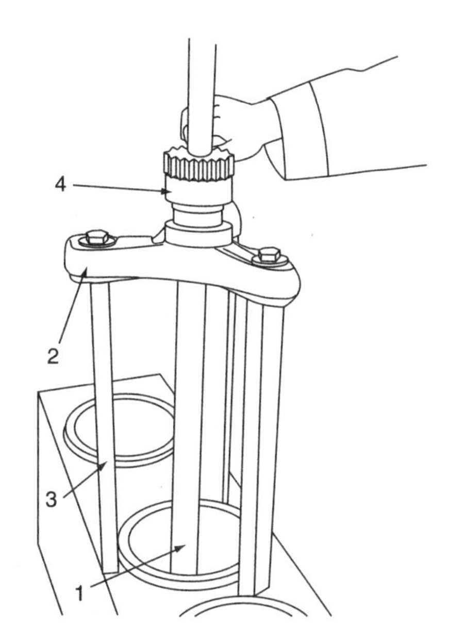

85. As shown in the figure above, Technician A says adapter plates must fit snugly in sleeves to prevent cocking in the block. Technician B says if sleeve puller plates are too large they are better because they cannot slip. Who is right?
    A. A only
    B. B only
    C. Both A and B
    D. Neither A nor B (C7)

86. Technician A says that a one to two ohm resistance between grounds is normal. Technician B says that electronic module's grounds can be checked with any standard volt/ohmmeter (VOM). Who is right?
    A. A only
    B. B only
    C. Both A and B
    D. Neither A nor B (F2.12)
87. Technician A says any generic scan tool can be used to read diagnostic trouble codes if the manufacturer's manual listed codes are used. Technician B says only the OEM manufacturer's scan tool can be used to perform approved reprogramming with the appropriate software cartridge. Who is right?
    A. A only
    B. B only
    C. Both A and B
    D. Neither A nor B (F2.12)
88. Technician A says to use a straightedge across the exhaust manifold surfaces to align separate in-line heads on the engine block. Technician B says to blow out head bolt holds to ensure bolts do not bottom out prematurely. Who is right?
    A. A only
    B. B only
    C. Both A and B
    D. Neither A nor B (E6)
89. A medium duty diesel engine with an electronic DS style distributor-type fuel injection pump needs a base timing adjustment. Technician A says you adjust the stepper motor on the side of the injection pump to change the timing. Technician B says that timing is adjusted using the OEM scan tool. Who is right?
    A. A only
    B. B only
    C. Both A and B
    D. Neither A nor B (F2.3)

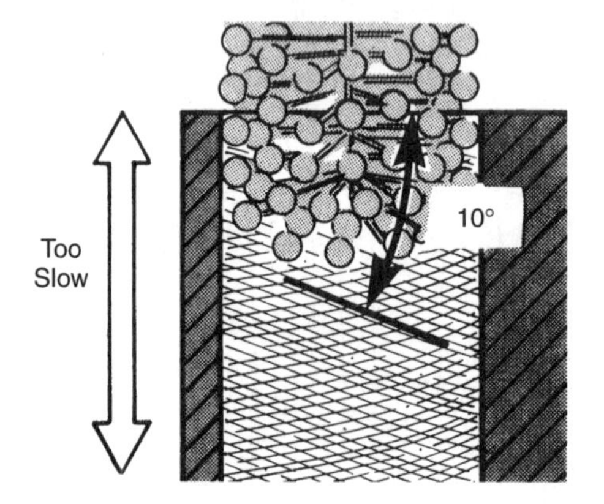

90. When using the tool in the figure above to hone a cylinder liner, Technician A says to never leave it in one position for too long. Technician B says when it strikes a high spot in the cylinder, there should be an increase in drag on the drill. Who is right?
    A. A only
    B. B only
    C. Both A and B
    D. Neither A nor B (C6)

91. When changing weather pack connectors the technician should be careful to:
    A. use the insertion/removal tool.
    B. bend the pins to fit.
    C. replace with standard connectors.
    D. ground all wiring before changing. (F2.13)
92. Technician A says that gaskets used on internal solenoids must be replaced with solenoids. Technician B says that care should be taken to note positioning of insulating washers. Who is right?
    A. A only
    B. B only
    C. Both A and B
    D. Neither A nor B (F2.6)
93. Technician A says bent push rods and tubes should be replaced, not straightened. Technician B says push rods or tubes must be checked for proper fit to cam followers. Who is right?
    A. A only
    B. B only
    C. Both A and B
    D. Neither A nor B (B15)
94. When performing a fuel injector pump lever (FIPL) test, the technician must:
    A. wait at least 10 minutes for codes to be issued from the scan tool.
    B. complete the test within 10 minutes of connecting the scan tool.
    C. have the engine running to perform this test.
    D. use a second technician to indicate when settings are too high or low. (F2.5)
95. Technician A says if the turbocharger lines and hoses become frayed or nicked they must be replaced, not repaired. Technician B says turbocharger oil tubes must be checked for obstructions. Who is right?
    A. A only
    B. B only
    C. Both A and B
    D. Neither A nor B (E2)
96. When replacing antifreeze, engine manufacturers recommend:
    A. ethylene glycol.
    B. methoxy propanol.
    C. either ethylene glycol and methoxy propanol, but not a mixture of both.
    D. A mixture of ethylene glycol and methoxy propanol. (D9)
97. Technician A says the hydraulic lifter body should be inspected for scuffing and scoring. Technician B says if the hydraulic lifter body is damaged, inspect the mating bore. Who is right?
    A. A only
    B. B only
    C. Both A and B
    D. Neither A nor B (B16)
98. Technician A says that when two or more identical belts are used on the same pulley, change only the belt that is bad. Technician B says belts can be installed by starting them on the edge and cranking the engine. Who is right?
    A. A only
    B. B only
    C. Both A and B
    D. Neither A nor B (D6)

99. Technician A says air intake suppressors are used on normally aspirated engines that use a Jake brake. Technician B says air intake suppressors are usually installed close to engine in the air intake system. Who is right?
    A. A only
    B. B only
    C. Both A and B
    D. Neither A nor B
    (H3)

100. Which component of a unit injector system is LEAST likely to be the cause of an improper injector spray pattern?
    A. Plugged nozzle tip holes
    B. A stuck injection nozzle
    C. Improper timing of injectors
    D. Excess injector-to-follower clearance
    (F1.10)

101. Technician A says an ammeter should be used to check for a short circuit between circuits. Technician B says to make sure you fully charge the battery before checking a circuit for current draw. Who is right?
    A. Technician A only
    B. Technician B only
    C. Both A and B
    D. Neither A nor B
    (G7)

102. Technician A says that petroleum jelly should be used on battery cable clamps to prevent corrosion. Technician B says to always use protective pads and a little grease to prevent corrosion on battery cable clamps. Who is right?
    A. Technician A only
    B. Technician B only
    C. Both A and B
    D. Neither A nor B
    (G5)

103. Technician A says that excessive smoke from a diesel engine indicates the engine needs to be overhauled. Technician B says that by discussing the engine operation with the operator you may find that an engine overhaul is not needed. Who is right?
    A. A only
    B. B only
    C. Both A and B
    D. Neither A nor B
    (A1)

104. A camshaft fails to pass visual inspection. Technician A says bad camshaft bearing surfaces can be turned. Technician B says flat spots on lobes can be silver soldered and remilled to specification. Who is right?
    A. A only
    B. B only
    C. Both A and B
    D. Neither A nor B
    (B18 and C8)

6 Additional Test Questions for Practice

Additional Test Questions

Please note the letter and number in parentheses following each question. They match the overview in section 4 that discusses the relevant subject matter. You may want to refer to the overview using this cross-referencing key to help with questions posing problems for you.

1. You suspect a cylinder head is warped, cracked, or has a blown head gasket. What would you do first?
   A. Check head bolt torque.
   B. Apply a dye crack detector to reveal the presence of small cracks.
   C. Remove all residue oil, carbon, and gasket material from surfaces and visually inspect for cracks.
   D. Remove excess scale deposits from the passage and perform a pressure test of the head to check for internal cracks. (B3)
2. Technician A says a radiator cap in a cooling system that cannot maintain a system pressure of its rating of 15 psi can cause an overheat condition. Technician B says when the pressure exceeds the rated amount the cap releases the pressure. Who is right?
   A. A only
   B. B only
   C. Both A and B
   D. Neither A nor B (D12)
3. Which procedure is LEAST likely to be used when performing valve seats maintenance?
   A. Testing seats for looseness with a ball peen hammer
   B. Sharpening a chisel to remove defective seats
   C. Checking seats with a concentricity indicator
   D. Dressing a valve seat grinding stone (B9)
4. A camshaft fails to pass visual inspection. Technician A says you cannot turn bad camshaft bearing surfaces. Technician B says you visually inspect the camshaft gear for cracks,chips or broken teeth. Who is right?
   A. A only
   B. B only
   C. Both A and B
   D. Neither A nor B (B18 and C8)
5. Which tool combination is needed to check valve head height?
   A. Straightedge and feeler gauge
   B. Dial indicator, step block, steel ruler, and outside caliper
   C. Vernier micrometer and point gauge
   D. Steel ruler and T-square (B3)

Additional Test Questions for Practice

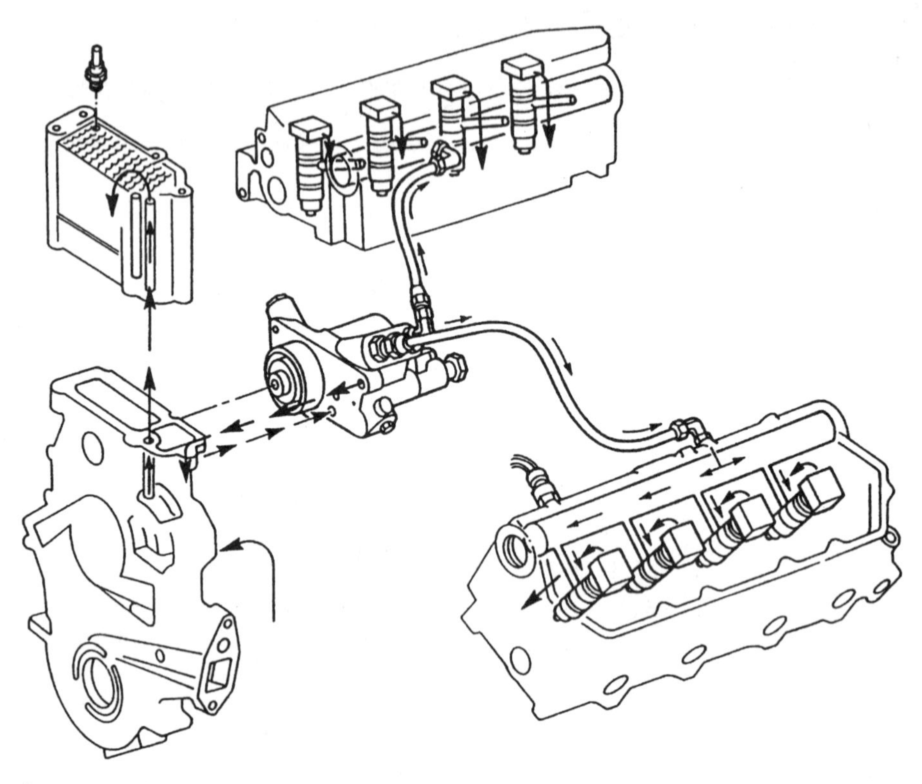

6. A electronically controlled common rail fuel system (HEUI) shown in the figure above has a low power complaint. Technician A says that the pump shown is supplying fuel to the injectors for combustion that could be the cause. Technician B says the high-pressure pump having low pressure could be the cause. Who is right?
   A. A only
   B. B only
   C. Both A and B
   D. Neither A nor B

   (F2.6)

7. When the governor is unstable what is the result?
   A. Under run
   B. High idle
   C. Hunting
   D. Low idle

   (F1.14)

8. Technician A says when boosting a vehicle with a discharged battery, always start the good vehicle before making connections so the battery will have enough cranking power. Technician B says to make the connections first so a short will not blow the good vehicle's charging system. Who is right?
   A. A only
   B. B only
   C. Both A and B
   D. Neither A nor B

   (G4)

9. Multiple fractures are discovered in the crankshaft. Technician A says the cause may be found by carefully inspecting the vibration damper. Technician B says the flywheel or torque converter needs to be closely inspected as a possible cause. Who is right?
   A. A only
   B. B only
   C. Both A and B
   D. Neither A nor B

   (C19)

10. When checking exhaust backpressure, Technician A says to connect a manometer to the exhaust manifold and check against manufacturer's specifications. Technician B says if a plug has not been provided for this test, drill and tap the exhaust manifold pipe. Who is right?
    A. A only
    B. B only
    C. Both A and B
    D. Neither A nor B (A9)

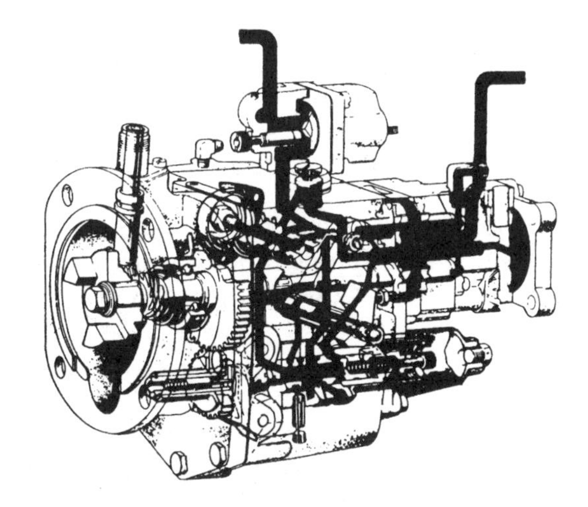

11. The PTG AFC pump shown in the figure above is an improvement over the standard PTG pump because it incorporates a(n):
    A. governor control.
    B. flow-no-flow by-pass valve.
    C. hydraulic clutch assembly.
    D. dual action pre-pump. (F1.8)
12. Which of the following is LEAST likely to happen?
    A. A faulty injector may cause an engine to produce a white smoke.
    B. Faulty or improperly installed oil control rings may result in the engine producing a white smoke.
    C. A blown head gasket may result in the engine producing either a blue or white smoke, depending on where the gasket is blown.
    D. Operating a diesel engine for extended periods in a low-load or no-load condition may result in the engine producing a black smoke or soot. (A5)
13. The timing adjustment on a distributor-type fuel injection pump uses which of the following processes:
    A. using a timing pin.
    B. using the spill method.
    C. adjusting the rack.
    D. rotating the pump. (F1.6)
14. An air intake is being checked for a restriction using a water manometer. Technician A says that if the air restriction is too excessive the filter or intake tubes need to be inspected. Technician B says that the engine speed should be the same for both readings. Who is right?
    A. A only
    B. B only
    C. Both A and B
    D. Neither A nor B (A7)

15. Technician A says to test an engine for excess crankcase pressure the engine must be at an idle. Technician B says the engine must be tested across a range of operating speeds. Who is right?
    A. A only
    B. B only
    C. Both A and B
    D. Neither A nor B (A10)
16. Technician A says coolant may form a pool of green or blue liquid below the component that is leaking. Technician B says that a coolant leak must leave some visible residue in the engine compartment or under the engine. Who is right?
    A. A only
    B. B only
    C. Both A and B
    D. Neither A nor B (A14)
17. Technician A says pistons should be laid out in the order in which they came out. Technician B says installing an improper injector nozzle could cause scoring and cracking of pistons skirts. Who is right?
    A. A only
    B. B only
    C. Both A and B
    D. Neither A nor B (A14)
18. Technician A says if valves are to be reused, do not grind valve stem ends as this will cause a change in tappet clearance. Technician B says existing valves should only be resurfaced if the valve seat was replaced. Who is right?
    A. A only
    B. B only
    C. Both A and B
    D. Neither A nor B (B8)
19. To perform a maximum delivery check on a Cummins PT pump, Technician A says that the torque screw and high idle screws must be backed out about 5 turns. Technician B says the pump must rotate to produce 1,000 strokes to collect the proper amount of fuel. Who is right?
    A. A only
    B. B only
    C. Both A and B
    D. Neither A nor B (F1.8)

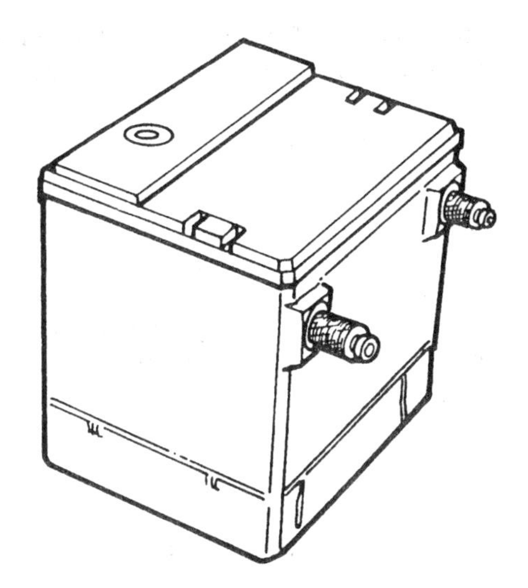

20. If severe corrosion is detected on cable and posts of a battery shown in the figure above, which is LEAST likely to be replaced?
    A. Battery posts
    B. Battery
    C. Battery cables
    D. Battery clamps (G5)
21. Technician A says most oil pump by-pass valves are factory sealed devices and should not be tampered with. Technician B says if an oil pump by-pass valve fails, the whole oil pump must be replaced. Who is right?
    A. A only
    B. B only
    C. Both A and B
    D. Neither A nor B (D3)
22. To perform the fuel injector pump lever (FIPL) sensor adjustments on a distributor injector pump:
    A. the key must be on and the engine running.
    B. the key must be on and the engine off.
    C. the key must be off and the engine off.
    D. the key must be off and the diagnostic tool set to engine run. (F2/3)
23. When the electric shutoff solenoid of an injector pump de-energizes, the return spring will:
    A. close a valve in the fuel line.
    B. push governor linkage to the no fuel position.
    C. dump the governor oil.
    D. bleed the high pressure fuel from the injector lines. (F1/6)

24. When installing starter, which is LEAST likely to be done?
    A. Connect control wiring to solenoid
    B. Align starter motor with flywheel
    C. Install starter mounting bolts
    D. Replace starter bushings (G9)
25. Technician A says that timing of an in-line injector pump and injection system may require replacement of shims between the cam and injector. Technician B says timing at the injector pump is accomplished by a camshaft inside the pump. Who is right?
    A. A only
    B. B only
    C. Both A and B
    D. Neither A nor B (F1.7)
26. All of the following are used to test battery capacity **EXCEPT:**
    A. multimeter set to milliamps.
    B. ammeter.
    C. voltmeter.
    D. carbon pile load. (G2)
27. Low engine power is a result of insufficient fuel delivery to a unit injector. Which of the following would be the LEAST likely cause?
    A. Suction leaks in fuel supply system
    B. Incorrect throttle travel
    C. Clogged fuel filter
    D. A leak in the injector return line (A12 and F1.9)

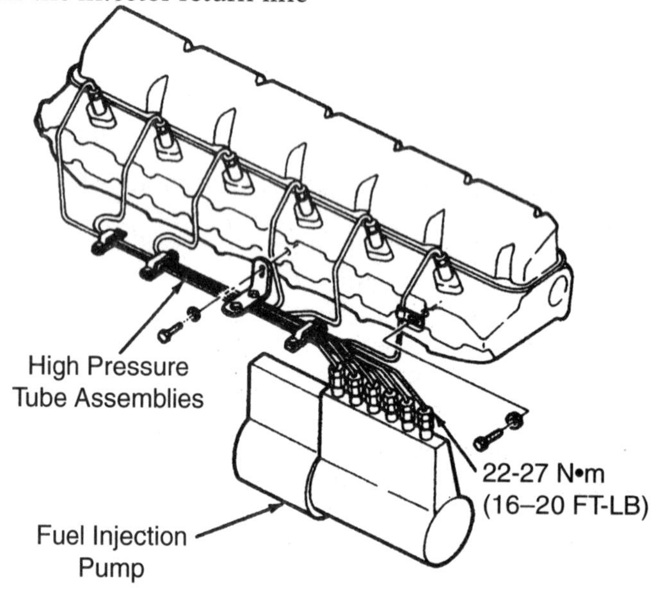

28. The injector pump system is being discussed in the figure above. Technician A says the lines shown are high-pressure fuel lines. Technician B says all fuel connections must be capped when they are disconnected. Who is right?
    A. A only
    B. B only
    C. Both A and B
    D. Neither A nor B (F1.12)

29. Cracks and scoring in the piston skirts are LEAST likely to be caused by which of the following?
    A. Excessive lubrication levels in the crankcase
    B. Improper piston clearance
    C. Excessive fuel setting
    D. Engine overheating (C17)
30. Which of the following is LEAST likely to be a correct connection of either the intake or exhaust system?
    A. Turbine outlet to exhaust manifold
    B. Compressor outlet to aftercooler inlet
    C. Aftercooler outlet to intake manifold
    D. Air filter to compressor inlet (E6)
31. Technician A says to replace valve springs that are fractured or broken, out of square or do not meet the specifications. Technician B says companion springs under the same valve bridge as defective springs must also be replaced. Who is right?
    A. A only
    B. B only
    C. Both A and B
    D. Neither A nor B (B5)
32. Technician A says "lugging" is when the throttle is in full fuel position and cannot pick up speed under load. Technician B says more fuel is burned than needed in the lugged condition. Who is right?
    A. A only
    B. B only
    C. Both A and B
    D. Neither A nor B (A1)
33. Technician A says when removing low-pressure fuel lines ensure the ground cable is removed from the battery. Technician B says line connections must be capped to prevent dirt from entering the fuel system. Who is right?
    A. A only
    B. B only
    C. Both A and B
    D. Neither A nor B (F1.13)

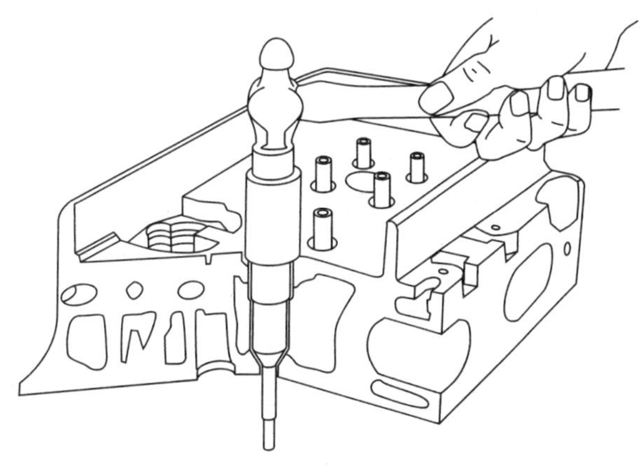

34. After replacing injector sleeves and seals as shown in the figure above, Technician A says injector tip height must be checked to confirm proper clearance. Technician B says to perform a cylinder head leak test. Who is right?
    A. A only
    B. B only
    C. Both A and B
    D. Neither A nor B (B11)

35. Which of the following is LEAST likely to provide satisfactory results when using sound to detect engine problems?
    A. Listening to the exhaust to determine which cylinder is misfiring.
    B. Using a bar-like stethoscope to find an open exhaust valve.
    C. Manually fouling out each injector and listening for a no change in sound condition.
    D. Selecting potentially faulty components based on the frequency of a sound in relation to the speed at which the component operates. (A4)
36. Technician A says that the customer should not disconnect the battery before seeking a service center when an emission control problem is detected. Technician B says that after clearing DTCs by disconnecting the battery, it may be necessary to operate the vehicle several times to resume normal operation. Who is right?
    A. A only
    B. B only
    C. Both A and B
    D. Neither A nor B (A1)
37. All of the following are acronyms associated with retrieving faults from a vehicle **EXCEPT:**
    A. (MIL) Malfunction Indicator Lamp.
    B. (DTC) Diagnostic Trouble Code
    C. (OBD II) On-Board Diagnostic II.
    D. (SBDS) Single Bit Diagnostic Subroutine. (A16)

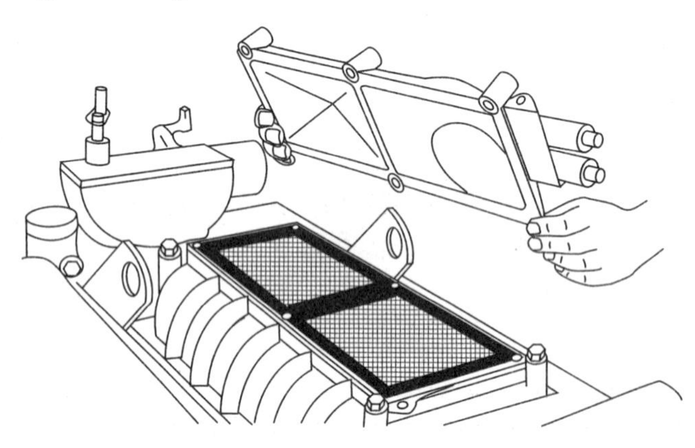

38. In the type blower shown in the figure above, Technician A says that continual use of the air shutdown valve while the engine is under heavy loads or at high rpm will result in damage to blower oil seals. Technician B says before replacing lightly scored blower rotor blades try to dress them with a file. Who is right?
    A. A only
    B. B only
    C. Both A and B
    D. Neither A nor B (E3)
39. When an installed engine oil pressure gauge is compared to a test gauge at low engine rpm, the installed gauge appears low, but at normal engine rpm, the installed gauge readings are correct. Technician A says that the oil passage in the block is clogged. Technician B says the line has a leak. Who is right?
    A. A only
    B. B only
    C. Both A and B
    D. Neither A nor B (A15)

40. When installing a new starter relay all of the following steps are required **EXCEPT:**
    A. install positive lead to relay.
    B. install negative lead to relay.
    C. install control wires on relay.
    D. install ground on battery. (G6)
41. Technician A says that the battery term ampere hour refers to stored charge capacity of a battery. Technician B says that a 75-ampere hour charge applied to a 200-ampere hour battery should turn the charge indicator green. Who is right?
    A. A only
    B. B only
    C. Both A and B
    D. Neither A nor B (G3)
42. Which tool is most likely to be used to determine the starter circuit voltage drop test?
    A. Starter shunt resistor
    B. Voltmeter
    C. Ohmmeter
    D. Milliammeter (G7)

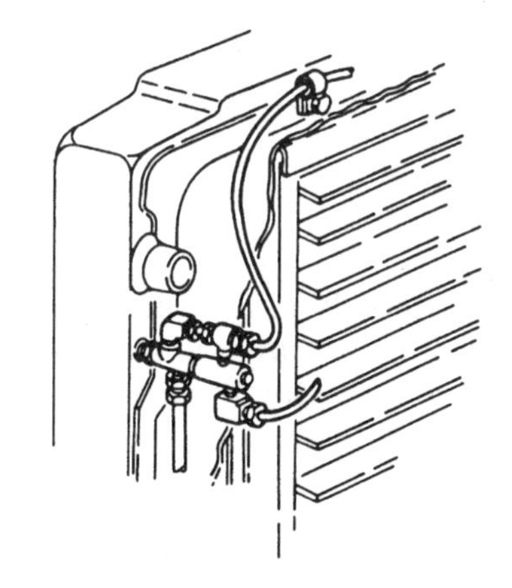

43. A thermostat piston controlled shutter, as shown in the figure above, relies on all of the following to regulate shutter operation **EXCEPT:**
    A. linkage adjustments.
    B. ambient air temperature.
    C. radiator temperature.
    D. air pressure. (D14)
44. A connector is found to be distorted from heat but when checked with an ohmmeter the electrical connection is still good. Technician A says to replace the connector. Technician B says to apply electrical tape around the connector. Who is right?
    A. A only
    B. B only
    C. Both A and B
    D. Neither A nor B (F2.13)
45. All of the following are true if a pressure cap is found to be defective **EXCEPT:**
    A. replace the cap.
    B. excess pressure could cause damage to the radiator tubes.
    C. excess vacuum pressure could collapse radiator hoses.
    D. adjust or replace the spring in the cap. (D12)

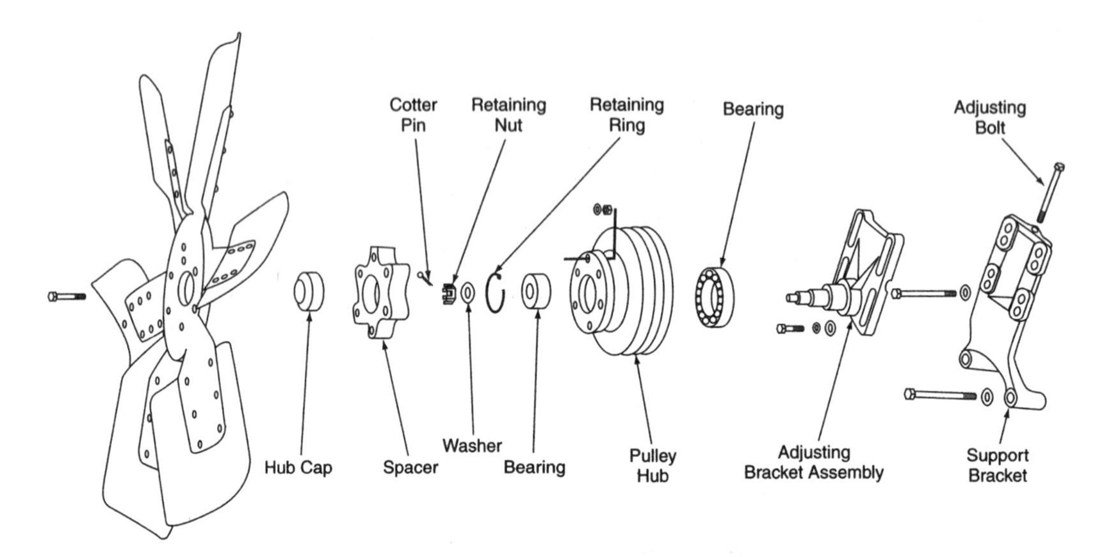

46. In the fan assembly parts in the figure above, Technician A says that fan bearings should be submerged in cleaning solvent and blown out with air to remove dirt. Technician B says because fan bearings are water cooled some corrosion is acceptable, although if corrosion is excessive they should be replaced. Who is right?
    A. A only
    B. B only
    C. Both A and B
    D. Neither A nor B (D13)

47. Technician A says canister filters should be pre-filled to protect the bearings. Technician B says that the filter housing mounted by-pass valve needs to be removed and inspected when replacing the filter. Who is right?
    A. A only
    B. B only
    C. Both A and B
    D. Neither A nor B (D3)

48. When the glow plug indicator light illuminates continuously, the:
    A. glow plug system is operational.
    B. glow plug system has experienced a short circuit.
    C. glow plugs are operating on a non-GM system.
    D. outside temperature requires driver to turn on the glow plug system. (E7)

49. Technician A says end gaps on rings must be offset to prevent blowby. Technician B says when installing a piston and connection rod, studs should be covered to prevent damage to the crankshaft. Who is right?
    A. A only
    B. B only
    C. Both A and B
    D. Neither A nor B (C16)

50. Droplets of water or gray oil on the dipstick would LEAST likely be caused by a:
    A. cracked block or cylinder head.
    B. blown head gasket.
    C. leaky oil cooler.
    D. worn oil pump. (A2)

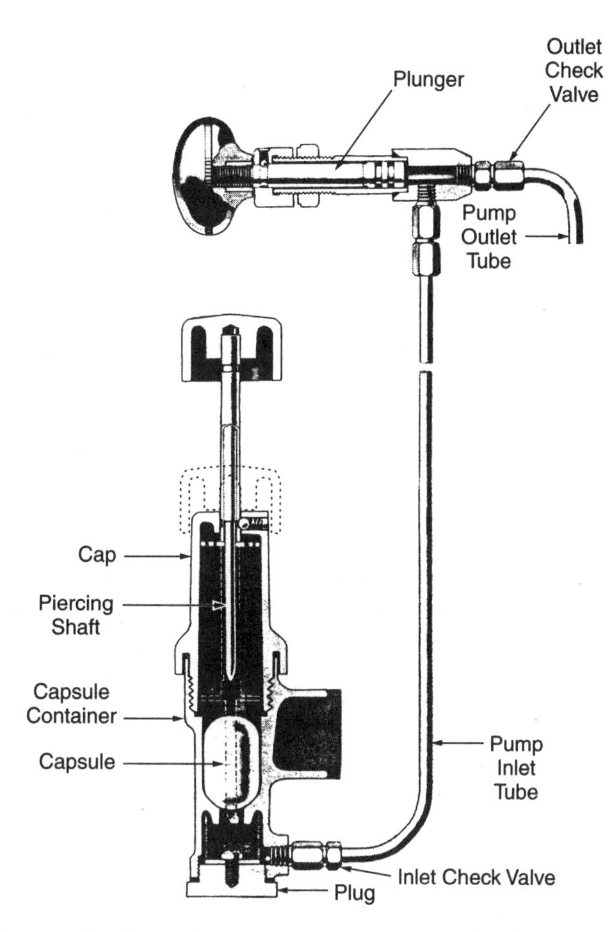

51. In the capsule-type fuel starting systems shown in the figure above, Technician A says check valves must be disassembled and cleaned and damaged balls and springs replaced. Technician B says plunger seals are not removable so the complete plunger assembly must be replaced. Who is right?
    A. A only
    B. B only
    C. Both A and B
    D. Neither A nor B (E8)
52. Technician A says valve rotators can be tested using a plastic mallet. Technician B says defective valve seals will cause oil consumption. Who is right?
    A. A only
    B. B only
    C. Both A and B
    D. Neither A nor B (B6)
53. Technician A says that crosshead guide pins should be at right angles to the heads milled surface. Technician B says to check crosshead guide pin diameter with a micrometer, and compare to manufacturer's specifications. Who is right?
    A. A only
    B. B only
    C. Both A and B
    D. Neither A nor B (B13)
54. To maintain an effective cooing system, all of the following must be done **EXCEPT:**
    A. the radiator and engine must be back-flushed annually.
    B. coolant removed during maintenance should be tested for corrosive elements.
    C. air must be bled from the engine during refilling of system.
    D. coolant should be tested for antifreeze protection level. (A14)

55. When replacing valve guides, Technician A says when the manufacture does not supply specifications, press the new guide in head to the same position the old guide was in. Technician B says excess scoring of the guide bore will require the head to be replaced. Who is right?
    A. A only
    B. B only
    C. Both A and B
    D. Neither A nor B (B7)

56. Technician A says that all gears in the timing geartrain must be inspected for tooth wear, including the crankshaft gear. Technician B says a slight roll or lip on each gear tooth is acceptable, because gears will normally mate this way during the first 500 miles of operation. Who is right?
    A. A only
    B. B only
    C. Both A and B
    D. Neither A nor B (C13)

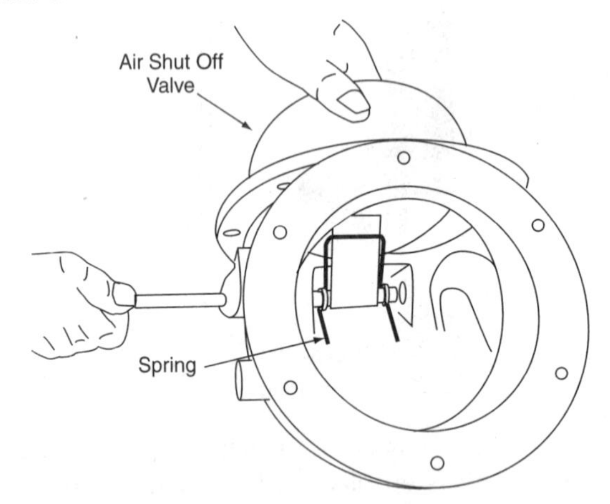

57. After removal and disassembly of the emergency shutdown device shown in the figure above all the following service procedures are appropriate **EXCEPT:**
    A. inspect parts for wear and replace as necessary.
    B. resurface face areas that are not flat.
    C. replace warped valve plates.
    D. do not replace blower screens. (E9)

58. Technician A says before installing die cast rocker covers on a cylinder head, ensure the silicone gasket is secure in the cover groove to ensure a good seal. Technician B says to always use new gaskets when installing valve covers. Who is right?
    A. A only
    B. B only
    C. Both A and B
    D. Neither A nor B (C1)

59. Technician A says that manufacturers allow for movement in the intake system components by using durable rubber couplings, which are clamped at both ends. Technician B says that if an air intake pipe, tube, or hose is loose or missing a clamp an inspection should be made of the turbocharger compressor and valves. Who is right?
    A. A only
    B. B only
    C. Both A and B
    D. Neither A nor B (E1 and E4)

60. When testing a cylinder head for potential coolant leakage, Technician A says the coolant needs to be heated to operating temperature. Technician B says the coolant needs to be under pressure. Who is right?
    A. A only
    B. B only
    C. Both A and B
    D. Neither A nor B (B4)
61. The control solenoid on an electronic fuel injector is being removed. Technician A says to remove the fuel shutoff solenoid, disconnect wire from terminals, remove the ball adjuster from the control rack, and remove the screws holding the solenoid. Technician B says when reinstalling, you position the solenoid and mount it using screws, then connect the ball adjuster to the rack and attach the wires. Who is right?
    A. A only
    B. B only
    C. Both A and B
    D. Neither A nor B (F2.15)
62. When pressure testing an engine block, Technician A says the block need only be submerged in water for about 20–30 minutes and the water checked for bubbles. Technician B says cracked cylinder blocks should be replaced. Who is right?
    A. A only
    B. B only
    C. Both A and B
    D. Neither A nor B (C3)

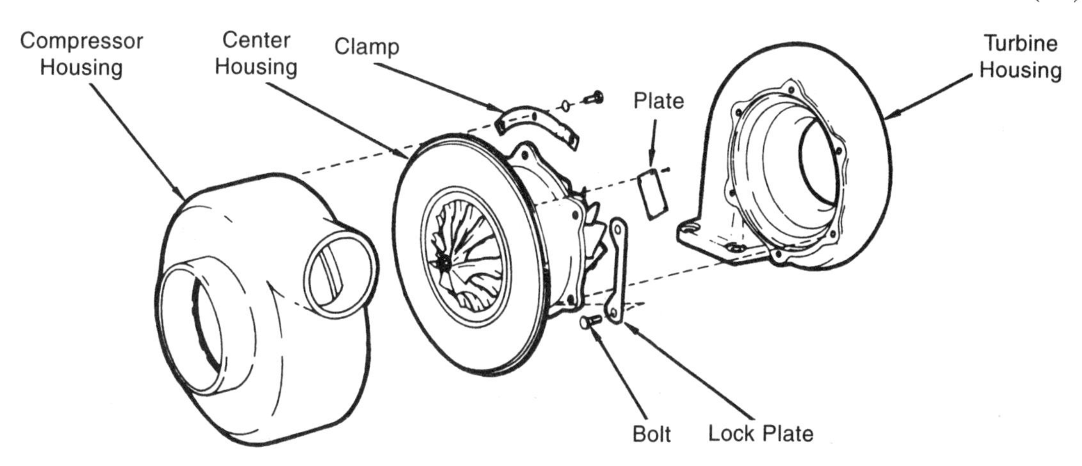

63. Technician A says that extreme caution is a must when working around operational turbochargers, as shown in the figure above, because the turbine turns between 60,000 to 100,000 rpm. Technician B says that a compressor turbine wheel must be replaced if it has a nicked or cracked blade. Who is right?
    A. A only
    B. B only
    C. Both A and B
    D. Neither A nor B (E2)
64. Which of the following is LEAST likely to be the cause of an engine not starting?
    A. The exhaust pipe is filled with soot or mud
    B. A short or open in the starter armature
    C. Corrosion on battery terminals
    D. An empty ether start canister (A11)

65. A blower air box is being tested for pressure. Technician A says to use a manometer to check pressure. Technician B says the engine must be operated at idle speeds. Who is right?
    A. A only
    B. B only
    C. Both A and B
    D. Neither A nor B (A8)

66. Technician A says that main bearing's shells should receive a light coat of oil before final installation. Technician B says that when checking the bearing, install a plastigage in all the bearings, torque to specifications, and rotate the crankshaft. Who is right?
    A. A only
    B. B only
    C. Both A and B
    D. Neither A nor B (C12)

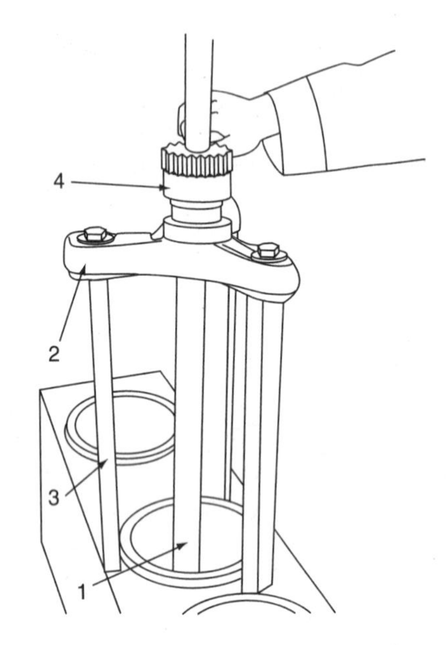

67. Technician A says the tool as shown in the figure above is used to bore out cylinder liners. Technician B says it is a cylinder sleeve puller. Who is right?
    A. A only
    B. B only
    C. Both A and B
    D. Neither A nor B (C7)

68. Technician A says the oil in the bottom chamber of an oil bath air cleaner is to lubricate air intake components. Technician B says this type air filter has a reduced efficiency at low rpm. Who is right?
    A. A only
    B. B only
    C. Both A and B
    D. Neither A nor B (E1)

69. Technician A says before installation head bolts must be cleaned and inspected for erosion or pitting. Technician B says the cylinder needs to be checked for foreign objects before installing the head. Who is right?
    A. A only
    B. B only
    C. Both A and B
    D. Neither A nor B (B1 and B2)

70. The single acting plunger-type fuel supply pump can be equipped with all of the following **EXCEPT:**
    A. sediment bowl.
    B. sediment strainer.
    C. hand primer.
    D. pop-off valve. (F1.2)
71. Technician A says when honing a cylinder liner the hone should shake strongly to provide a good cleaning. Technician B says once the honing has been started do not remove it, even to clean the stones. Who is right?
    A. A only
    B. B only
    C. Both A and B
    D. Neither A nor B (C6)

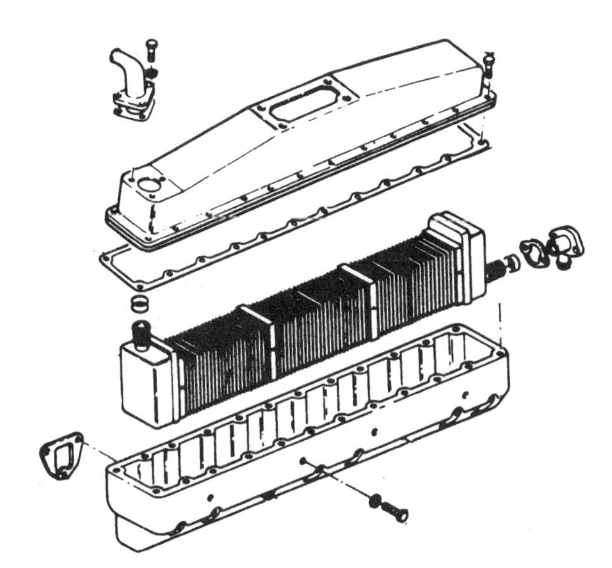

72. Reduced power and high coolant temperature are noted on an engine with an intercooler, as shown in the figure above. Technician A says you need to bleed the air out of the intercooler. Technician B says the air tubes in the cooler may be clogged. Who is right?
    A. A only
    B. B only
    C. Both A and B
    D. Neither A nor B (E5)
73. All of the following must be checked before installing rocker arms, **EXCEPT:**
    A. rocker arm bushings for wear.
    B. rocker arm shaft for wear.
    C. rocker arm shaft for pitting or scoring.
    D. rocker arm tappet screws are tightened down. (B15)
74. Technician A says small leaks can easily be detected, because antifreeze does not evaporate and generally leaves a red or green colored buildup or stain nearby. Technician B says a defective thermostat is most frequently caused by repeated overheating of the engine. Who is right?
    A. A only
    B. B only
    C. Both A and B
    D. Neither A nor B (D8)

75. During the overhaul of an engine a broken crank is discovered. Technician A says the operator was hot-riding the engine. Technician B says an out-of-round condition may exist on one or more of the main bores. Who is right?
    A. A only
    B. B only
    C. Both A and B
    D. Neither A nor B (C10)
76. Technician A says gauges work better than indicators because you can see when they are not working. Technician B says failure of a digital or electronic gauge may not be the fault of the actual gauge circuit. Who is right?
    A. A only
    B. B only
    C. Both A and B
    D. Neither A nor B (F1.16)
77. Technician A says that only a manufacturer's scan tool is compatible with diagnostic data link protocol. Technician B says that each manufacturer uses a different data link connector, so only their scan tool can be used. Who is right?
    A. A only
    B. B only
    C. Both A and B
    D. Neither A nor B (F2.14)
78. Technician A says to use a digital voltmeter to test grounds with power applied. Technician B says to use an ammeter to test grounds when power is not applied. Who is right?
    A. A only
    B. B only
    C. Both A and B
    D. Neither A nor B (F2.12)

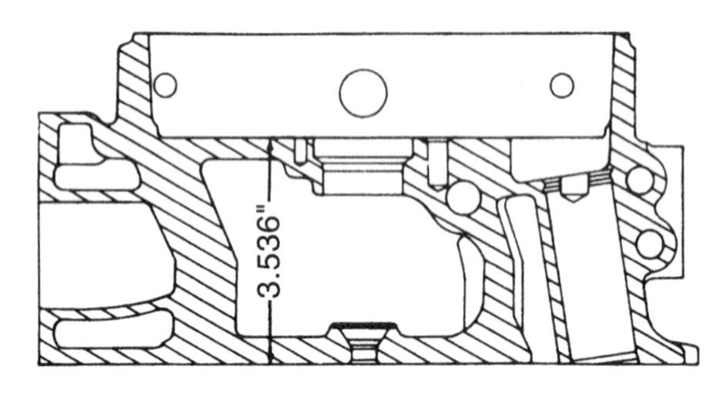

79. The cylinder head deck-to-deck thickness measurement indicated in the figure above is used to:
    A. prevent cylinder head warpage.
    B. determine valve guide height.
    C. determine the extent of cylinder warpage.
    D. determine the maximum allowable resurfacing. (B10)
80. When the throttle is moved from wide open to closed the throttle position sensor should indicate:
    A. 0 VDC to 5 VDC.
    B. 5 VDC to 0 VDC.
    C. 5 VDC to continuous.
    D. 0 VDC to continuous. (F2.2)

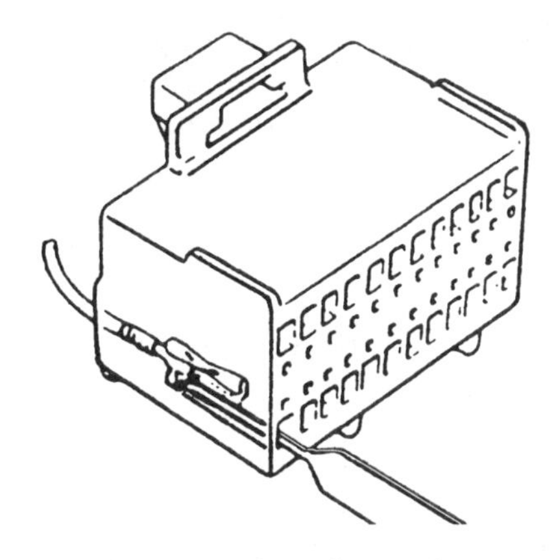

81. The figure above indicates the procedure for replacing which type of connector?
    A. Micro
    B. Weather pack
    C. Pull-to-seal
    D. Molded plug (F2.13)
82. Technician A says small leaks can easily be detected because antifreeze does not evaporate and generally leaves a red or green colored buildup or stain nearby. Technician B says a defective thermostat is most frequently caused by repeated over-heating of the engine. Who is right?
    A. A only
    B. B only
    C. Both A
    D. Neithe (D8 and D9)

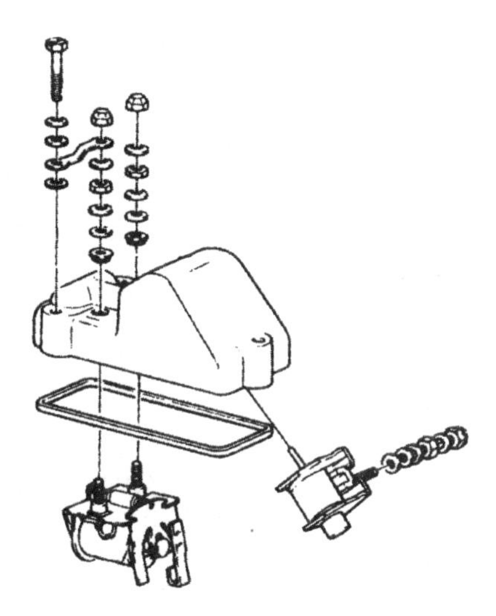

83. In the internal shutdown solenoid in the figure above, Technician A says that it must be tested after replacement. Technician B says to clean the old gasket off the cover and injection pump when changing the shutdown solenoid. Who is right?
    A. A only
    B. B only
    C. Both A and B
    D. Neither A nor B (F1.6)

84. Technician A says a precision oil pressure test gauge should mimic the instrument panel gauge in readings. Technician B says test readings should be taken during start up, varying operating ranges, and on shut down. Who is right?
    A. A only
    B. B only
    C. Both A and B
    D. Neither A nor B (A15)
85. Which of the following is LEAST likely to produce a loss of power in a diesel at full rated loads?
    A. High exhaust back pressure
    B. High density air supplied to the cylinder
    C. Oil and dirt in the air tubes or cooling fins of the aftercooler
    D. Small leak in the fuel line from the tank to the pump, or a crimped line to an injector (A12)
86. The throttle position sensor is being adjusted on a fuel injection pump with electronic controls. Technician A says that on some applications, a gauge block is placed between the gauge boss and the maximum throttle travel screw. Technician B says the scan tool will signal when the setting is too low or too high. Who is right?
    A. A only
    B. B only
    C. Both A and B
    D. Neither A nor B (F2/3)
87. Technician A says ethylene glycol and methoxy propanol should not be mixed. Technician B says if switching the type of antifreeze used, the engine must be flushed completely first. Who is right?
    A. A only
    B. B only
    C. Both A and B
    D. Neither A nor B (D9)
88. Technician A says that oil in a hold-down bolt hole of a head can throw off the torque setting of that bolt. Technician B says that a head bolt with stripped threads should be redressed before installing. Who is right?
    A. A only
    B. B only
    C. Both A and B
    D. Neither A nor B (C2)
89. Technician A says wet cylinder sleeves should be cleaned by immersion or with glass bead plating and then closely inspected. Technician B says a cylinder out-of-round condition can be detected by measuring at one point and rotating around the inside 180 degrees. Who is right?
    A. A only
    B. B only
    C. Both A and B
    D. Neither A nor B (C4)

Voltage and Temperature Chart

TEMPERATURE	MINIMUM VOLTAGE
21°C (70°F & Above)	9.6
10°C (50°F)	9.4
–1°C (30°F)	9.1
–10°C (15°F)	8.8
–18°C (0°F)	8.5

90. As shown in the chart above, what would be the minimum allowable voltage drop during a battery load test if the temperature is 50°F (10°C)?
    A. 9.6
    B. 9.4
    C. 9.1
    D. 9.8 (G2)
91. A small crack is detected in a wet-sleeve counterbore. Technician A says to use a heliarc to seal the crack and re-cut the counterbore. Technician B says the block needs to be sent out to have the counterbore re-sleeved. Who is right?
    A. A only
    B. B only
    C. Both A and B
    D. Neither A nor B (C4)
92. Technician A says a lifter foot should be smooth and slightly concave. Technician B says if a lifter foot is pitted, the camshaft mating lobe should be inspected. Who is right?
    A. A only
    B. B only
    C. Both A and B
    D. Neither A nor B (B16)
93. An engine has a vibration occurrence rate that is equal to 1/2 the engine rpm. Technician A says that this might be caused by a bent cam. Technician B says that this could be caused by a bent connecting rod. Who is right?
    A. A only
    B. B only
    C. Both A and B
    D. Neither A nor B (A13)
94. Technician A says alignment between pulleys must not exceed 1/16 in. (1.59 mm) for each 12 in. (30.5 cm) of distance between pulley centers. Technician B says belts must not touch the bottom of grooves and protrude more than 3/32 in. (2.38 mm) above the top of the groove. Who is right?
    A. A only
    B. B only
    C. Both A and B
    D. Neither A nor B (D11)
95. Technician A says that testing of the oil cooler is to be done between engine overhauls. Technician B says that testing of the oil cooler is to be done during an engine overhaul. Who is right?
    A. A only
    B. B only
    C. Both A and B
    D. Neither A nor B (D4)

96. To eliminate pulsation in the air intake system caused by exhaust brakes install:
    A. an air intake suppressor.
    B. a wastegate.
    C. an oil bath air filter.
    D. a variable camshaft. (H3)

97. Technician A says that PID codes are retrieved by using the diagnostic test set. Technician B says that PID codes can be cleared by removing the negative battery cable. Who is right?
    A. A only
    B. B only
    C. Both A and B
    D. Neither A nor B (F2.1)

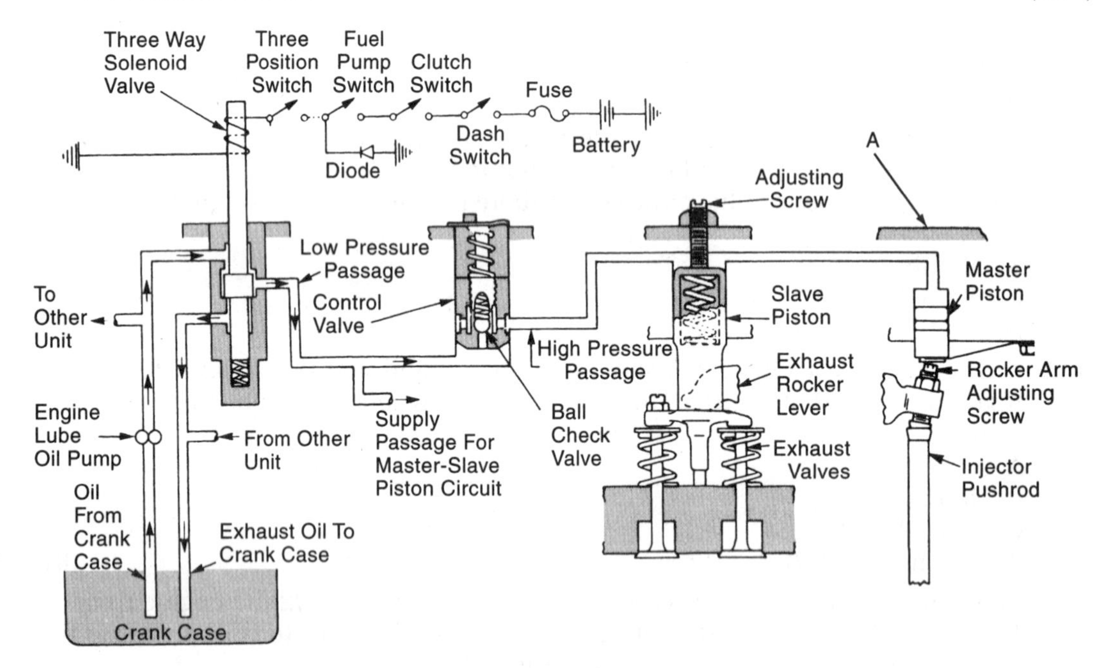

98. As shown in the figure above, the diode in the circuit is used to:
    A. apply the ground to the battery.
    B. rectify the signal applied to the battery.
    C. shunt the stored potential when the coil is de-energized.
    D. provide a radio frequency interference shunt. (H2)

99. As shown in the figure above question 98, the device at position A must:
    A. activate the injector below it.
    B. sense the push rod upward motion to open exhaust valves.
    C. sense the push rod downward motion to open exhaust valves.
    D. push the rocker arm below it to open the intake valves. (H1)

100. Technician A says in a PT injector system when the pressure of fuel against the angular surface of the injector needle exceeds the spring tension, the injector fires. Technician B says in a unit injector system when the cam lobe presses the plunger of the injector down, it pressurizes fuel and injects fuel into the cylinder. Who is right?
    A. A only
    B. B only
    C. Both A and B
    D. Neither A nor B (F1.8)

101. A two-stoke diesel engine is being tuned. Which of the adjustments listed should be done first?
   A. Adjust the valves
   B. Adjust the low speed spring
   C. Adjust the governor gap
   D. Balance the valve bridges (B17)
102. A precombustion chamber is being replaced on a medium duty diesel engine. Which of these processes apply in this repair?
   A. Press the prechamber in the cylinder head with an arbor press.
   B. Adjust the clearance using a grinding wheel.
   C. Check the clearance with a feeler gauge flush to plus .002 in.
   D. Use 400 grit sandpaper to grind the prechamber to fit. (B12)
103. While discussing a computer controlled diesel engine, Technician A says corrosion on wire connectors can affect engine performance. Technician B says that after repairing wire connectors, they should be treated with a corrosion inhibitor. Who is right?
   A. A only
   B. B only
   C. Both A and B
   D. Neither A nor B (F2.13)
104. A knocking noise is heard in the engine. The noise increases at the same rate as engine rpm. Which of the following is LEAST likely to cause this?
   A. Fan belt
   B. Camshaft
   C. Crankshaft
   D. Rocker arms (A4)
105. Technician A says over-fueling can cause very black smoke to come out of the exhaust pipe. Technician B says that the coolant level can cause this problem. Who is right?
   A. A only
   B. B only
   C. Both A and B
   D. Neither A nor B (A5)
106. When replacing the powertrain control module (PCM), on an engine with electronic unit injectors, what must the technician do first?
   A. Disconnect the battery
   B. Connect a diagnostic tool to the vehicle
   C. Remove injector lines
   D. Turn the key on (F2.7)

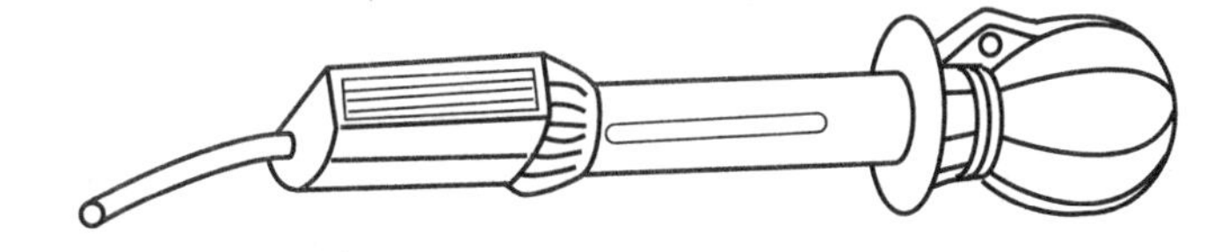

107. The tool shown in the figure above is used to check the specific gravity of the electrolyte in the battery. What is the name of this tool?
   A. Refractometer
   B. SCA
   C. Hydrometer
   D. Thermometer (D10)

108. While discussing a low voltage problem at the starter, Technician A says that a worn or shorted battery cable may cause this. Technician B says that a bad starter drive may cause this. Who is right?
A. A only
B. B only
C. Both A and B
D. Neither A nor B (G6)

109. When removing the starter from a vehicle what should the technician remove first?
A. The starter bolts
B. The starter solenoid
C. The negative battery cable
D. The inspection cover (G9)

110. The injector pump and pump drive on a diesel engine may be lubricated by all of the following **EXCEPT:**
A. lubricity of the diesel fuel.
B. directly from the oil pump in the engine.
C. from an oil port in the engine block.
D. from the transfer pump. (F1.11)

111. Technician A says to start a diesel engine in very cold weather you may use starter fluid. Technician B says an immersion heater in the oil or coolant will help start the engine easier. Who is right?
A. A only
B. B only
C. Both A and B
D. Neither A nor B (E8)

112. A diesel engine that has a power loss or rough operating problem can be caused by all of the following **EXCEPT:**
A. a plugged air filter.
B. a leaking fuel line.
C. a plugged fuel filter.
D. a low coolant level. (A12)

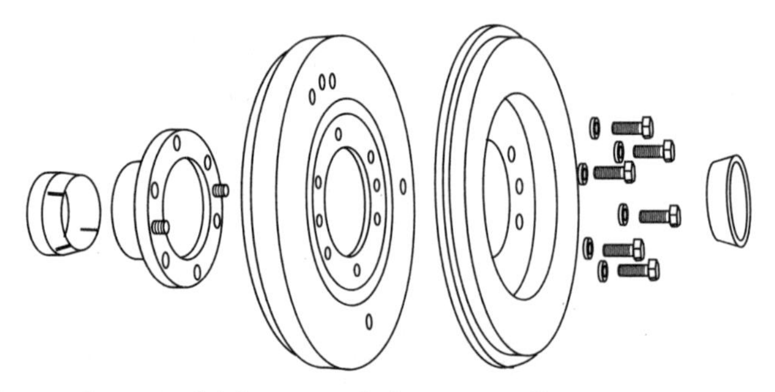

113. The above figure shows which part of the engine?
A. Flywheel assembly
B. Vibration damper assembly
C. Timing gear assembly
D. Water pump assembly (C19)

114. Technician A says if the pressure in the radiator is too great, the pressure cap will release the excess pressure and fluid into an expansion tank. Technician B says most radiator pressure caps have a release pressure of 15 psi or lower. Who is right?
A. A only
B. B only
C. Both A and B
D. Neither A nor B (D12)

115. While checking the oil level in the engine the technician notices coolant in the engine oil. All of the following could cause this **EXCEPT:**
    A. a crack in the exhaust valve.
    B. a crack in the water jacket in the engine block.
    C. an internal leak in the water pump.
    D. a leak in the head gasket. (D11)
116. On a vehicle with the two battery 12-volt system, the battery's connection is which one of the following?
    A. Series circuit
    B. Parallel circuit
    C. DC circuit
    D. AC circuit (G8)

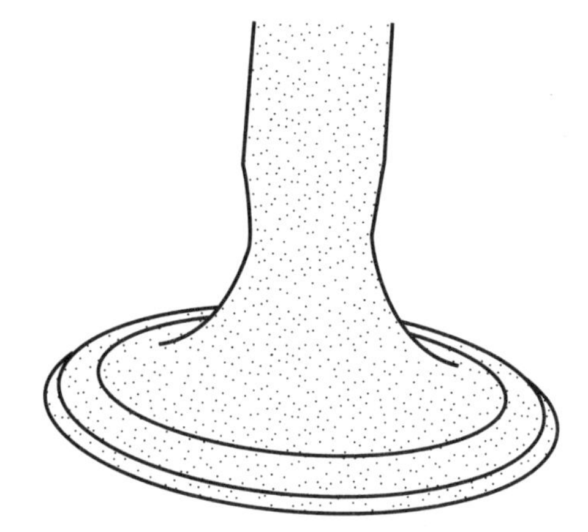

117. The above figure shows which of the following valve conditions?
    A. Cracks in the valve stem
    B. Carbon deposits on the face of the valve
    C. Cracks on the face of the valve
    D. A stretched valve stem (B8)
118. Which of the following should the technician check first when reinstalling a cylinder head to the engine block?
    A. Holes for foreign matter and cross threading
    B. The water pump bearings
    C. The front crank seal
    D. The rear crank seal (B14)
119. Technician A says the cylinder head should be checked for warpage. Technician B says that the wrist pins should be checked for wear. Who is right?
    A. A only.
    B. B only.
    C. Both A and B.
    D. Neither A nor B. (C18)
120. All of the following could cause fuel to contaminate the engine oil **EXCEPT:**
    A. a cracked fuel pot in the head.
    B. a broken piston ring.
    C. a leaking valve cover gasket.
    D. a leaking injector seal. (A15)
121. Which of the following is LEAST likely to be used to check the inside diameter of a valve guide?
    A. Snap gauge
    B. Feeler gauge
    C. Ball gauge
    D. Dial indicator (B7)

122. Technician A says that mechanical rotators on a valve spring will help eliminate carbon deposits on the valve seat. Technician B says that they can prevent oil leakage from the valve guide. Who is right?
   A. A only
   B. B only
   C. Both A and B
   D. Neither A nor B (B6)

123. All of the following will indicate valve seat leaks **EXCEPT:**
   A. visible cracks.
   B. oblonged valve seats.
   C. discoloration of the back of the valve head.
   D. worn piston rings. (B9)

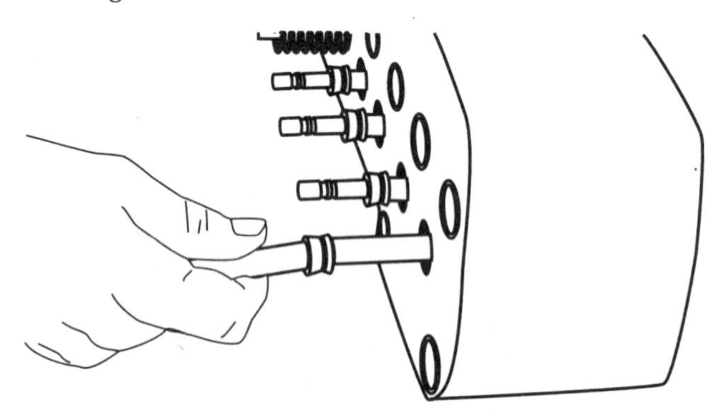

124. As shown in the figure above, a technician is performing what procedure on the cylinder head?
   A. Replacing valve guide seals
   B. Replacing valve guides
   C. Replacing valve stems
   D. Replacing rocker arm studs (B6)

125. The removal of the injector from a cylinder head is being discussed. Technician A says you use a slide hammer to remove the injector. Technician B says you can also use a puller to remove the injector. Who is right?
   A. A only
   B. B only
   C. Both A and B
   D. Neither A nor B (F1.10 and F2.6)

126. Before installing the valves into the cylinder head you should check which one of the following first?
   A. Valve to seat contact
   B. Rocker arm adjustment
   C. Lift on the camshaft
   D. Head gasket (B14)

127. Technician A says that with a wet sleeve type cylinder, the cylinder only seals at the top. Technician B says in this type of liner the coolant can pass around the cylinder. Who is right?
   A. A only
   B. B only
   C. Both A and B
   D. Neither A nor B (C7)

128. When reinstalling a cylinder sleeve all of the following should be checked **EXCEPT:**
    A. the mating surfaces between the sleeve and the cylinder block.
    B. the mating surfaces between the sleeve and O-rings.
    C. the inside of the sleeve for scoring.
    D. the clearance between the crankshaft and the main caps. (C5)
129. When removing a piston from the sleeve, which of these should the technician do first?
    A. Remove the cam shaft
    B. Remove the water pump
    C. Remove the top ridge
    D. Remove the sleeve seals (C6)

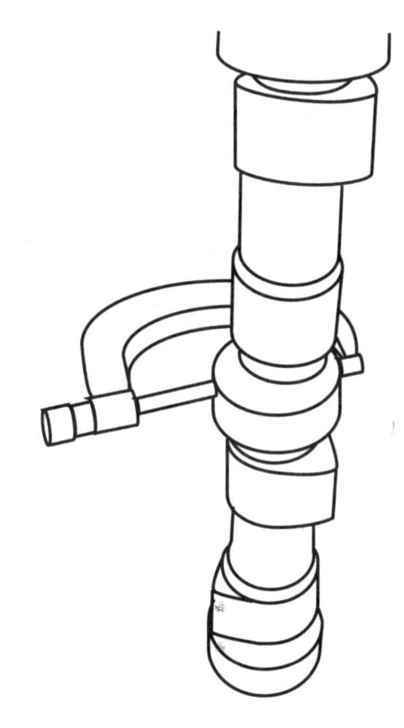

130. As shown in the figure above, what is the technician checking for?
    A. Cam lobe wear
    B. Cam bearing journal wear
    C. Cam lift
    D. Cam duration (B18 and C8)
131. Technician A says a micrometer is used to measure the crankshaft journals. Technician B says that when cleaning the journals you can use a wire wheel on a die grinder. Who is right?
    A. A only
    B. B only
    C. Both A and B
    D. Neither A nor B (C10)

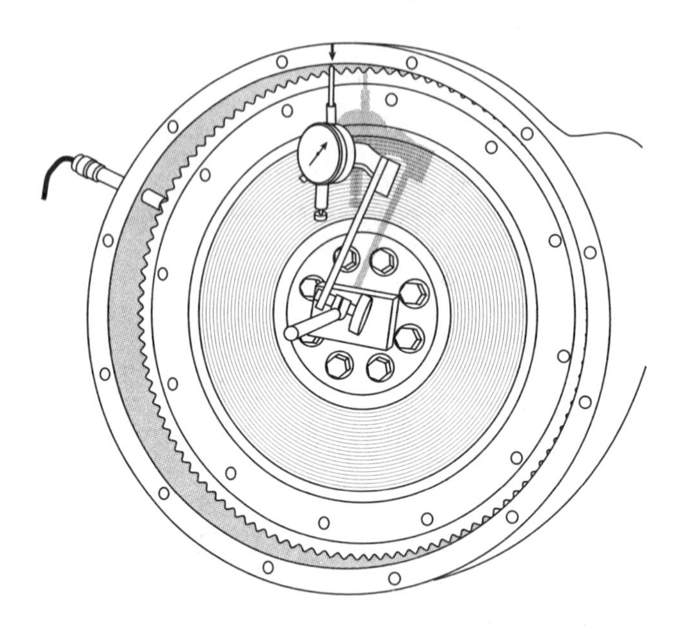

132. The setup shown in the figure above is used to check what on the flywheel housing:
    A. flywheel lateral runout.
    B. face lateral runout.
    C. crankshaft flange runout.
    D. housing bore radial runout. (C21)
133. Which of the following is LEAST likely to be used to measure the backlash in the timing gear train?
    A. Dial indicator
    B. Feeler gauge
    C. Depth gauge
    D. Ruler (C13)
134. Which of the following is LEAST likely to be a symptom of a governor problem?
    A. Low engine power
    B. High idle overrun
    C. Low idle underrun
    D. A rough running engine (F14)
135. Technician A says that it is easier to install piston rings on an aluminum piston if you preheat the piston in 200°F (93°C) water. Technician B says that the ring ends should be staggered. Who is right?
    A. A only
    B. B only
    C. Both A and B
    D. Neither A nor B (C16)

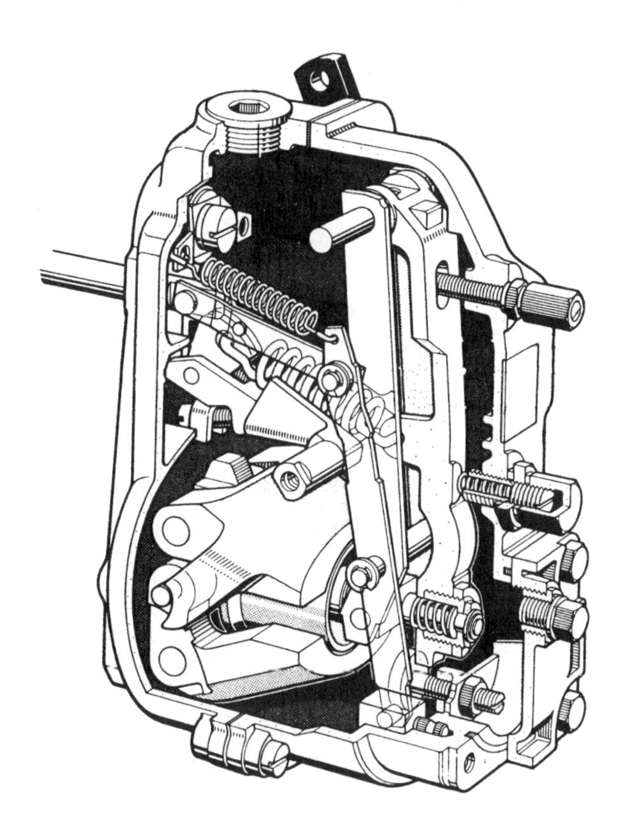

136. What type of governor is shown in the figure above?
    A. Hydraulic
    B. Mechanical
    C. Pneumatic
    D. Electrical (F1.14)
137. When removing the engine from the vehicle the technician should do which of the following first to ensure crankshaft balance?
    A. Check torque on the main bearings
    B. Mark the input shaft on the transmission
    C. Check the motor mounts
    D. Mark all drivetrain components for alignment before removal (C17)
138. When replacing the existing main bearing with oversize bearings, Technician A says the crankshaft journals should be machined to fit the bearings. Technician B says the main caps should also be line bored. Who is right?
    A. A only
    B. B only
    C. Both A and B
    D. Neither A nor B (C11)
139. On a vehicle with an electrical oil pressure gauge all of the following will cause the faulty readings **EXCEPT:**
    A. low voltage to the gauge.
    B. a short in the wire from the sending unit.
    C. low coolant level.
    D. a defective sending unit. (F1.16)

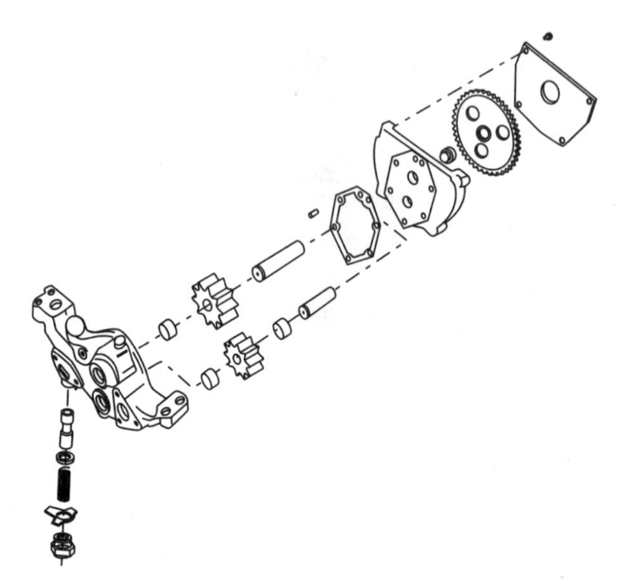

140. The figure above shows what type of oil pump?
    A. External
    B. Internal
    C. Electrical
    D. Pneumatic (D2)

141. Two technicians are discussing the lubrication system on a diesel engine. Technician A says that there are two different types of filters used, the cartridge and the spin-on type. Technician B says they also use a partial-type and a full-flow type. Who is right?
    A. A only
    B. B only
    C. Both A and B
    D. Neither A nor B (D3)

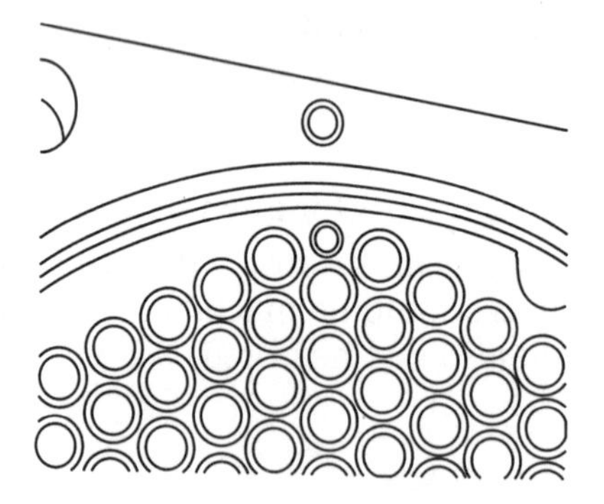

142. The figure shown above shows what part of the lubrication system?
    A. Oil filter
    B. Oil cooler
    C. Filter housing
    D. Oil pump (D4)

143. A technician suspects the mechanical temperature gauge is giving false readings. To check the temperature gauge all of the following could be used **EXCEPT:**
    A. heat sensitive tape.
    B. radiator thermometer.
    C. heat gun.
    D. voltmeter. (D7)

144. On a vehicle with an over cooling problem, Technician A says that a stuck thermostat could cause this problem. Technician B says that an overfilled radiator could cause this problem. Who is right?
    A. A only
    B. B only
    C. Both A and B
    D. Neither A nor B (A14)

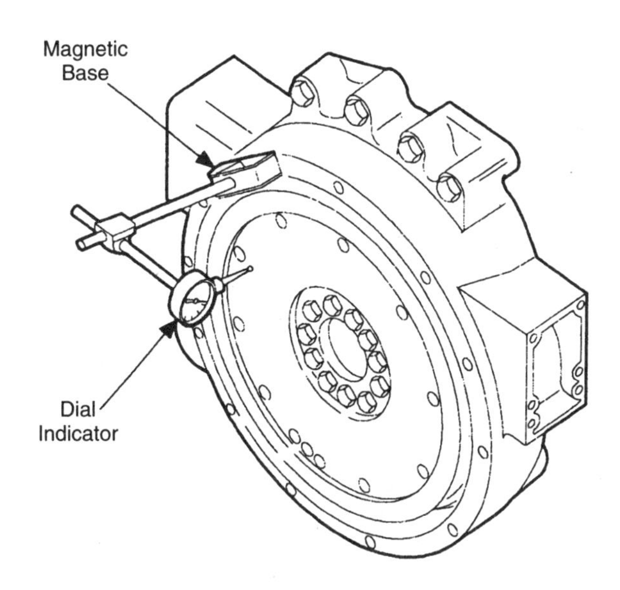

145. What flywheel measure process is being performed in the figure above?
    A. Clutch face surface runout
    B. Pilot bore radial runout
    C. Bore runout
    D. Bore concentricity (C21)
146. Technician A says the starter ring gear is usually pinned or bolted to the flywheel assembly. Technician B says the ring gear is a shrink fit to the flywheel. Who is right?
    A. A only
    B. B only
    C. Both A and B
    D. Neither A nor B (C21)
147. Technician A says that when installing an in-line injector pump with an electronic governor, it must be timed with the engine camshaft to ensure proper injection timing. Technician B says an internal camshaft in the pump determines the firing order. Who is right?
    A. A only
    B. B only
    C. Both A and B
    D. Neither A nor B (F2.4)
148. Which of the following is least likely to cause low fuel pressure?
    A. A partially plugged fuel filter
    B. Low engine idle
    C. A sticking injector
    D. Low coolant level (F1.4)

149. Technician A says that high-pressure fuel lines should be bled of any air that might be trapped after replacement. Technician B says you can check for air in the fuel system by installing a clear plastic hose in the return line and look for large bubbles. Who is right?
   A. A only
   B. B only
   C. Both A and B
   D. Neither A nor B (F1.3)

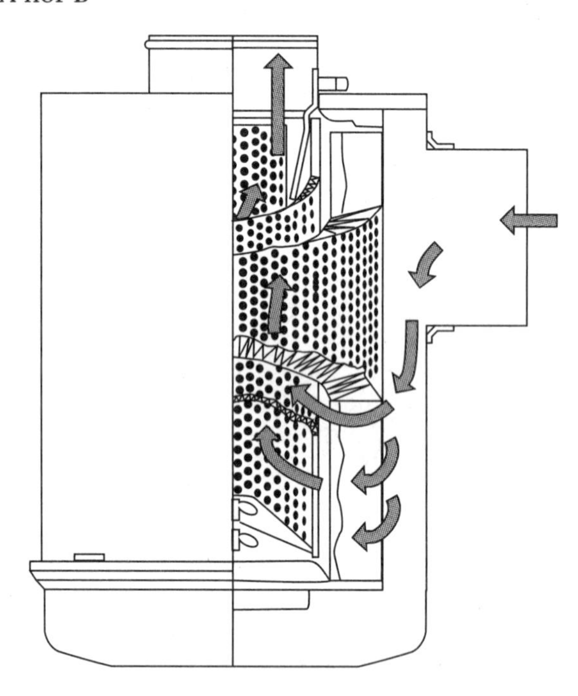

150. What type of air cleaner is shown in the figure above?
   A. Oil bath
   B. Dry type
   C. Centrifugal
   D. Aftercooler (E1)

151. All of the following are connected to the turbocharger **EXCEPT:**
   A. the air intake and air filter.
   B. the exhaust manifold.
   C. an oil feed line.
   D. the water pump. (E6)

152. While inspecting a rotor-type blower, the technician notices an oil film on the blower rotors. Which of the following is LEAST likely to cause this?
   A. Blower bearing seal leaks
   B. Worn rotor bearing
   C. Restriction in the oil feed line
   D. Worn end plate gaskets (E3)

153. While discussing a turbocharged diesel engine, Technician A says that an exhaust leak will affect the performance of the engine. Technician B says that leaks in the intake pipe can cause damage to the intake side of the turbo. Who is right?
   A. A only
   B. B only
   C. Both A and B
   D. Neither A nor B (A2)

154. On a diesel engine with an intercooler, the technician notices coolant in the intake side of the turbocharger. What should the technician check first?
   A. Intercooler
   B. Water pump
   C. Intake manifold
   D. Head gasket (E5)

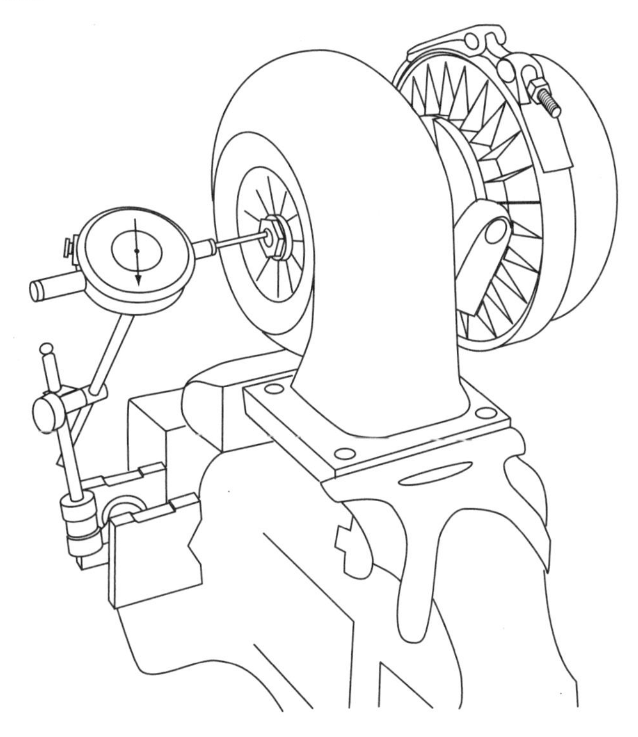

155. As shown in the figure above, the technician is checking the turbo for:
   A. clearance between the exhaust turbine and the turbo housing.
   B. end play in the turbine shaft.
   C. turbine fins for wear.
   D. checking the compressor fin for wear. (E2)

156. Technician A says that a compression brake will cause the engine to stop the vehicle immediately. Technician B says that the compression brake opens the exhaust valves before the compression stroke is complete. Who is right?
   A. A only
   B. B only
   C. Both A and B
   D. Neither A nor B (H2)

157. Which of the following is LEAST likely to be found in a compression brake system?
   A. Master piston
   B. Slave piston
   C. Solenoid valve
   D. Brake shoes (H1)

158. Under load a diesel engine emits a very dark black smoke from the exhaust pipe. What should the technician check first?
   A. Air cleaner
   B. Fuel pump
   C. Injector pump
   D. Turbocharger (A4)

159. Technician A says that low air box pressure can cause an air inlet restriction. Technician B says that you can check air box pressure with a water manometer. Who is right?
 A. A only
 B. B only
 C. Both A and B
 D. Neither A nor B
 (A8)
160. On a vehicle with an electric fuel shut off valve, which of the following is not part of this system?
 A. Oil pressure gauge
 B. Key switch
 C. Electric solenoid
 D. Governor linkage
 (F1.15)
161. While inspecting the exhaust system, the technician notices that the vehicle has an excess amount of exhaust pressure. Which of the following is LEAST likely to cause this?
 A. Improper muffler type
 B. Excessive carbon buildup in the exhaust system
 C. Collapsed exhaust pipe
 D. Low oil pressure
 (A9)
162. Technician A says that excessive crankcase pressure can be caused from a plugged oil breather. Technician B says that all diesel engines use EGR valves. Who is right?
 A. A only
 B. B only
 C. Both A and B
 D. Neither A nor B
 (A10)
163. When trying to erase the PID error codes without the diagnostic test set the technician can do which of the following to accomplish this?
 A. Turn off the key switch
 B. Disconnect the battery cable
 C. Unplug engine sensors
 D. Remove the electronic control module (ECM)
 (A16)
164. All of the following should be inspected when removing low pressure fuel lines **EXCEPT:**
 A. Any kinks in the lines
 B. Connectors on the lines
 C. Cracks in the lines
 D. Power steering pump
 (F1.13)
165. Technician A says if an engine has low horsepower, the fuel filter should be checked. Technician B says that the high idle setting could cause this also. Who is right?
 A. A only
 B. B only
 C. Both A and B
 D. Neither A nor B
 (F1.5)
166. Technician A says the powertrain control module (PCM) contains emission-related information codes. Technician B says you must have a scan tool to clear the emission codes. Who is right?
 A. A only
 B. B only
 C. Both A and B
 D. Neither A nor B
 (F2.1)

167. What is the meaning of the abbreviation "DTC" as it pertains to the error codes on the modern diesel engine?
    A. Drive Train Codes
    B. Diagnostic Trouble Codes
    C. Direct Trouble Codes
    D. Direct Transmission Codes (F2.1)

168. When removing a starter from the vehicle, what should the technician do first?
    A. Remove the battery
    B. Remove the starter solenoid
    C. Disconnect the voltage regulator
    D. Disconnect the battery cables (G9)

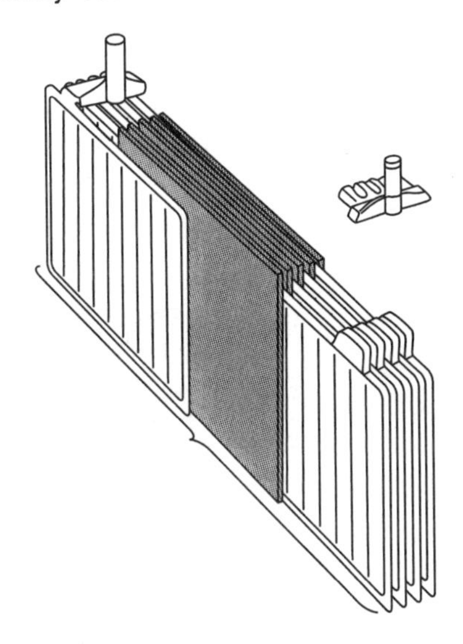

169. The figure shown above is what part of the battery?
    A. Side post types
    B. Positive and negative plates
    C. Top post type
    D. Bottom plates (G1)

170. Technician A says by using a pressure cap on the radiator, the boiling point of the water will be raised. Technician B says that pressure caps come with many different pressure settings. Who is right?
    A. A only
    B. B only
    C. Both A and B
    D. Neither A nor B (D4)

171. After replacing an injector line what will the technician have to do first before starting the engine?
    A. Readjust the low idle
    B. Bleed all air out of the line
    C. Readjust the high idle
    D. Replace the transfer pump (F1/4)

172. Technician A says that when performing a starter draw test you should not crank the engine for more than 30 seconds at a time. Technician B says you can perform this test with a battery charger connected to the battery. Who is right?
    A. A only
    B. B only
    C. Both A and B
    D. Neither A nor B (G7)

173. When performing a load test on a battery, the technician finds that the battery voltage drops below manufacturer's specifications. The technician should do which one of the following?
    A. Recharge the battery and return it to service
    B. Recharge the battery and retest it
    C. Replace the battery
    D. Replace the voltage regulator (G1)
174. Technician A says that radiator air shutters are used to help maintain coolant temperature. Technician B says that air shutters are used to keep out foreign debris from the radiator also. Who is right?
    A. A only
    B. B only
    C. Both A and B
    D. Neither A nor B (D14)

175. The figure above shows what type of fan hub?
    A. Air applied fan hub
    B. Electrically activated fan hub
    C. Vicious-clutch fan hub
    D. Hydraulic applied fan hub (D13)
176. Technician A says that glow plugs are used to help start the engine in cold weather. Technician B says that glow plugs should be disconnected in warm weather. Who is right?
    A. A only
    B. B only
    C. Both A and B
    D. Neither A nor B (E7)
177. All of the following are parts in a starting aid pump system **EXCEPT:**
    A. the pump body.
    B. the inlet check valve.
    C. the outlet check valve.
    D. the fan pulley. (E8)

178. Technician A says that the four valve heads on a four stroke engine use four exhaust valves in each cylinder. Technician B says that a four stroke diesel engine with valve crossheads will use a four valve head. Who is right?
   A. A only
   B. B only
   C. Both A and B
   D. Neither A nor B (B13)

179. All of the following can be used to bleed air out of the fuel system **EXCEPT:**
   A. by cranking the engine and bleeding the injector lines.
   B. by pressurizing the fuel system with the primer pump and bleeding the injector lines.
   C. by replacing the fuel filter.
   D. by pressurizing the fuel tank with a low volume air line and bleeding the injector lines. (F1.4)

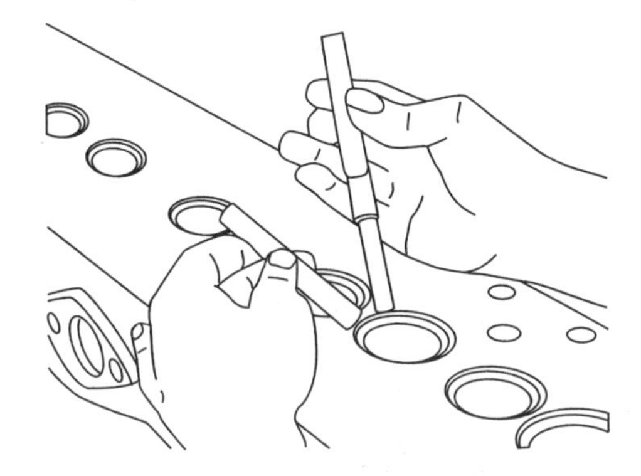

180. The figure above shows the technician is replacing which part of the head?
   A. Valve seat
   B. Valve guide
   C. Valve stem
   D. Valve guide seal (F10)

181. Which of the following is LEAST likely to have the air shut down valve mounted to it.
   A. Blower
   B. Inlet side of the turbo charger
   C. Intake manifold
   D. Exhaust manifold (E9)

182. All of the following are parts of a valve cover **EXCEPT:**
   A. the head rail.
   B. the breather housing.
   C. the filler cap.
   D. the strainer. (B1)

183. Technician A says that a shutoff solenoid has two coils in one housing. Technician B says that one coil pulls the fuel shutoff to the "open" position. Who is right?
   A. A only
   B. B only
   C. Both A and B
   D. Neither A nor B (F1.15)

184. After the technician cleans the engine block, what should he do first to the block?
   A. Paint the engine block
   B. Test the engine block for cracks
   C. Remove all cylinder sleeves
   D. Plug all oil ports (C3)

185. Technician A says that when replacing a wet-type sleeve you should always replace the seals on the sleeve. Technician B says that to remove the sleeve you should tap it out with a hammer. Who is right?
    A. A only
    B. B only
    C. Both A and B
    D. Neither A nor B (C7)

186. Which is LEAST likely to be used to check connections on the electronic modules?
    A. Digital ohmmeter
    B. Digital multimeter
    C. High impedance test light.
    D. Load tester (A3)

187. Technician A says that the customer option codes can be changed in the electronic systems with a scan tool. Technician B says in some cases changing these codes could result in poor engine performance. Who is right?
    A. A only
    B. B only
    C. Both A and B
    D. Neither A nor B (A1)

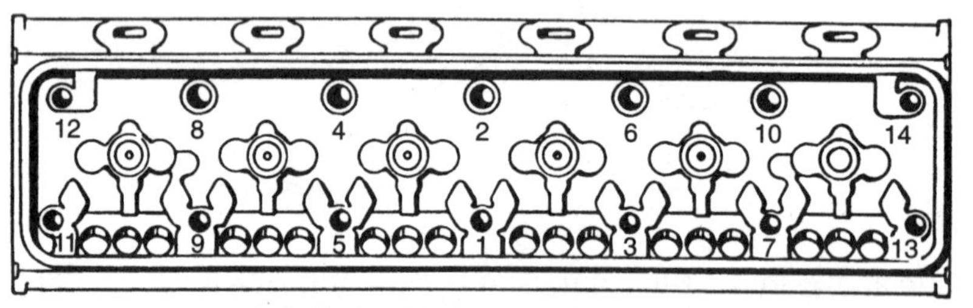

6-Cylinder Engine Cylinder Head

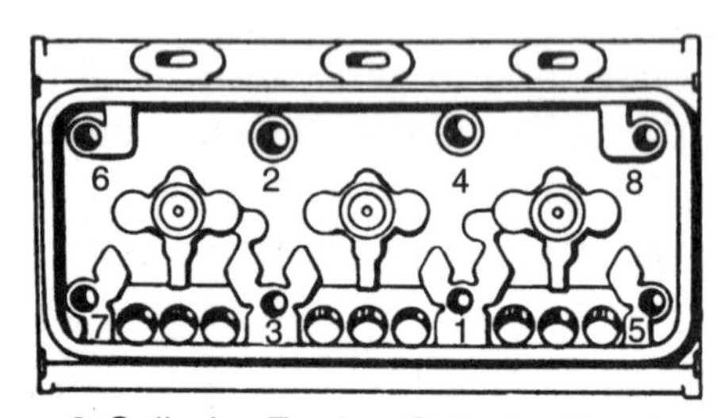

3-Cylinder Engine Cylinder Head

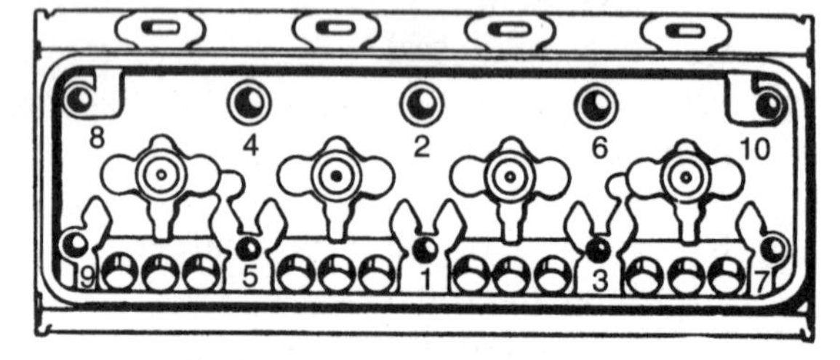

4-Cylinder Engine Cylinder Head

188. The figure shown above shows a sequence of numbers. These numbers are used for what purpose?
    A. To tell the technician which bolts to put in first
    B. To tell the technician which bolts to tighten first
    C. The serial number of the head
    D. The part number of the head (B14)

189. Technician A says that the throttle position sensor is connected to a 5-Volt reference. Technician B says when the throttle is manually opened, the scan tool should read a low voltage. Who is right?
    A. A only
    B. B only
    C. Both A and B
    D. Neither A nor B (F2.2)

190. Which of the following must the technician do first whenever the cylinder liners are removed?
   A. The main bearing caps must be inspected
   B. The block bores must be inspected
   C. The oil ports must be inspected
   D. The cam bearing must be inspected (C4)

191. Technician A says the EMC connectors should never be probed through the weather pack seals. Technician B says a test kit should be used to test connectors. Who is right?
   A. A only
   B. B only
   C. Both A and B
   D. Neither A nor B (F2.13)

192. When removing an internal shutdown solenoid, what must the technician remove first?
   A. The solenoid from the cover
   B. The pump cover
   C. The injector pump
   D. The air box (F2.10)

193. Technician A says cylinder heads are always torque in sequence according to specifications set by the manufacturer. Technician B says that any push rods that are bent must be straightened. Who is right?
   A. A only
   B. B only
   C. Both A and B
   D. Neither A nor B (B15)

194. Technician A says that on a turbo charged diesel engine, the turbo charger is lubricated by engine oil. Technician B says that if the turbo shaft bearing goes out, the engine oil must be changed. Who is right?
   A. A only
   B. B only
   C. Both A and B
   D. Neither A nor B (D5)

195. The required clearance between the rotor lobes on an engine driven blower is obtained by which of the following?
   A. Adding or removing shims
   B. Changing the end cover thrust plate
   C. Using over or undersized gears
   D. Using over or undersized bearings (E3)

196. Technician A says that when replacing the lifters in the engine you should also replace the timing gears. Technician B says that the lifter should be checked in the mating bore in the block to ensure proper fit. Who is right?
   A. A only
   B. B only
   C. Both A and B
   D. Neither A nor B (B16)

197. What would the technician do first when adjusting drive belts on an engine?
   A. Replace the drive pulleys
   B. Replace the idler pulleys
   C. Check belts for cracking and wear
   D. Apply belt dressing compound (D6)

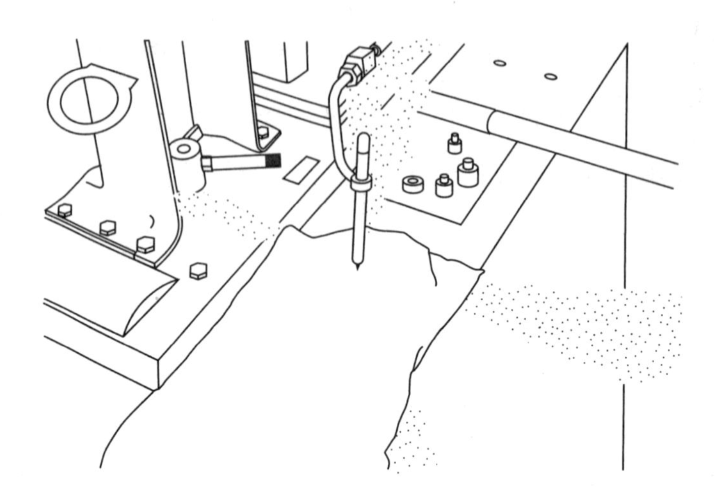

198. In the figure above, the technician is performing what test in the fuel injection system?
    A. Injector pump
    B. High-pressure lines
    C. Low-pressure lines
    D. Injector pattern

(F1.10)

7 Appendices

Answers to the Test Questions for the Sample Test Section 5

1.	A	27.	C	53.	A	79.	B
2.	C	28.	B	54.	C	80.	D
3.	D	29.	A	55.	B	81.	B
4.	C	30.	A	56.	C	82.	A
5.	B	31.	C	57.	A	83.	C
6.	D	32.	D	58.	D	84.	C
7.	B	33.	C	59.	C	85.	A
8.	B	34.	D	60.	C	86.	D
9.	C	35.	B	61.	C	87.	A
10.	B	36.	D	62.	A	88.	C
11.	D	37.	A	63.	A	89.	B
12.	C	38.	C	64.	C	90.	C
13.	A	39.	C	65.	A	91.	A
14.	B	40.	A	66.	B	92.	C
15.	A	41.	C	67.	C	93.	C
16.	A	42.	C	68.	B	94.	A
17.	A	43.	A	69.	C	95.	C
18.	B	44.	B	70.	C	96.	A
19.	A	45.	D	71.	C	97.	C
20.	D	46.	C	72.	D	98.	D
21.	A	47.	C	73.	B	99.	C
22.	B	48.	A	74.	B	100.	C
23.	D	49.	A	75.	A	101.	B
24.	D	50.	C	76.	D	102.	C
25.	D	51.	A	77.	D	103.	B
26.	A	52.	A	78.	C	104.	D

Explanations to the Answers for the Sample Test Section 5

Question #1
Answer A is correct because the figure shows a distributor pump that distributes fuel to nozzles.
Answer B is wrong because the figure does not indicate a jerk pump that is an indivdual pump for each cylinder.
Answer C is wrong because the figure does not indicate an in-line pump that contains a pumping element for each cylinder.
Answer D is wrong because the figure does not indicate a unit injector.

Question #2
Answer A is a good choice because a light, slippery, almost invisible coating on most of the engine compartment surfaces could be caused by a small leak in the radiator. Yet it is wrong because both technicians are right.
Answer B is also a good choice because this could be from excess venting of radiator pressure or a very small leak in a high-pressure fuel or hydraulic line. Yet it is also wrong because both technicians are right.
Answer C is correct because a light, slippery, almost invisible coating on most of the engine compartment surfaces could be caused by a small leak in the radiator. This could be from excess venting of radiator pressure or a very small leak in a high-pressure fuel or hydraulic line.
Answer D is wrong because both technicians are correct.

Question #3
Answer A is wrong because a loose or improperly mated connector can present a problem to the circuit and is a likely cause.
Answer B is wrong because wires that are nicked are exposed to the elements and will likely start to corrode. As the corrosion forces the insulation to expand, a lump will form in the wire.
Answer C is wrong because a melted or distorted electrical connector or wire is exposed to the elements and can likely become corroded. Electrical contact points and wires that have been overheated break down at the molecular level and corrosion starts within fractures in the structure.
Answer D is correct because ground wires attached to an unpainted frame surface make a positive contact and unless they become loose are least likely to cause a problem.

Question #4
Answer A is a good choice because you can use a metal bar like a stethoscope to find the bad cylinder but it is wrong because both technicians are right.
Answer B is also a good choice because you can manually foul each injector and listen for the one that does not have an affect on engine performance but it is wrong because both technicians are right.
Answer C is correct because both technicians are right.
Answer D is wrong because neither technician is wrong.

Question #5
Answer A is wrong because the smoke may appear white at first, but as the engine temperature increases it will turn black. This usually takes several minutes during which time the injector will dribble excess fuel into the cylinder. When the engine temperature increases enough for combustion this may cause precombustion or improper burning of fuel, hence the black smoke. A broken injector spring will cause dribble and black smoke after startup.
Answer B is correct because white smoke results from incomplete combustion in overlean combustion chamber areas, by fuel spray impingement on metal surfaces, and with low temperatures in the cylinder, such as when starting an engine (more so on cold days). It is like pouring oil on a hot exhaust pipe.
Answer C is wrong because Technician A is wrong.
Answer D is wrong because one of the technicians is right.

Question #6
Answer A is wrong because the diagnostic tool is used to give an audible indication when the setting is correct.
Answer B is wrong because you remove the two screws holding the sensor and rotate the sensor.
Answer C is wrong because the rotation of the sensor moves it in and out to calibrate the sensor for the lever position, not a tone sound.
Answer D is correct because you do not adjust the maximum throttle travel screw.

Question #7
Answer A is wrong because this will cause a short circuit and will damage both batteries.
Answer B is correct because a parallel circuit allows the maximum current flow.
Answer C is wrong because this connection can cause gasing and if batteries are gassing they may explode.
Answer D is wrong because if vehicles are making an electrical ground when the positive cable is attached, it might arc; if the battery is gassing, it could explode.

Question #8
Answer A is wrong because this will overload the fuse and could cause damage to the ohmmeter. Use a volt meter.
Answer B is correct because checking the voltage drop between (+) positive and the starter terminal stud while the starter is cranking is a voltage drop check on cable connection resistance.
Answer C is wrong because only Technician B is right.
Answer D is wrong because one of the technicians is right.

Question #9
Answer A is a good choice because the ambient engine temperature may be too low to obtain combustion temperature and start the engine, but it is wrong because both technicians are right.
Answer B is also a good choice because you need to check for a low battery or corrosion on the battery terminals and clean and charge as necessary, but it is also wrong because Technician A is also right.
Answer C is correct because both technicians are right.
Answer D is wrong because neither technician is wrong.

Question #10
Answer A is wrong because a multimeter is a combination of several measuring devices and the question asks for the specific measuring instruments.
Answer B is correct because you use a voltmeter and ammeter to measure battery voltage and capacity.
Answer C is wrong because only one technician is right.
Answer D is wrong because only one technician is right.

Question #11
Answer A is wrong because the ground cable is connected between the starter and ground.
Answer B is wrong because positive lead is connected to the starter.
Answer C is wrong because removal of the battery ground will prevent any electrical short.
Answer D is correct because it is not necessary to remove the battery in most starter removals, unless it is due to accessibility and in this figure this is not the case.

Question #12
Answer A is a good choice because high altitude or low fuel level on a grade can cause a random loss of power due to low combustion heat, but A is wrong because both technicians are right.
Answer B is also a good choice because a windchill on the tank may have reduced the fuel temperature to less than 10°F (6°C) of the fuel's pour point and cause random loss of power due to fuel starvation, but is also wrong because both technicians are right.
Answer C is correct because both technicians are right for the above reasons.
Answer D is wrong because neither technician is wrong.

Question #13
Answer A is correct because an out of balance turbocharger is the least likely cause of a vibration that is directly proportional to engine speed, because the turbine turns at speeds significantly higher than engine rpm.
Answer B is wrong because the vibration frequency would be low, but the rate of occurrence could be equal to the engine speed. The rod is probably hitting the wall at the same point with every rotation of the crankshaft and this is a likely cause.
Answer C is wrong because this type of vibration is likely to cause the complete engine to wobble on its mounts as the crank turns. The occurrence rate at any give point tested would appear to be equal to the engine rpm.
Answer D is wrong because the vibration damper is calibrated to the engine it is installed on, and is attached to the crank. Any change in the balance of the vibration damper will appear as a change in the overall balance of the crank and the vibration occurrence rate will be equal to the engine rpm and therefore a likely cause.

Question #14
Answer A is wrong because this would cause a higher-than-normal or over-temperature condition at high speed and load operation, generally leading to a constant overflow of fluid and formations of air pockets and hot spots. Indications of this condition would be most evident at the instrument panel until an air pocket formed around the temperature sensor(s), then the indication would drop to near zero.
Answer B is correct because a leaky fuel injector will not cause excessive surges in coolant level at normal temperatures.
Answer C is wrong because this condition would result in higher than normal coolant temperature after normal warmup. Progressive restrictions may lead to excessive cylinder head temp and then warping until it blows a head gasket.
Answer D is wrong because this condition will result in prolonged loss of fluid even when the engine is not hot.

Question #15
Answer A is correct because a clogged oil pressure relief valve creates a restriction resulting in higher pressures and is the least likely cause.
Answer B is wrong because if the screen or suction tube are restricted there will not be enough oil for the pump to deliver, so the pressure will be low.
Answer C is wrong because if the bearings are worn then it will take more oil to fill the space and the pump may not be able to maintain pressure.
Answer D is wrong because this will reduce the amount of oil flowing from the pump.

Question #16
Answer A is correct because the component shown is a tandem fuel pump.
Answer B is wrong because a check valve is not shown.
Answer C is wrong because a pressure regulator is not shown.
Answer D is wrong because a pressure sensor is not shown.

Question #17
Answer A is correct because this process is not necessary when checking for cracks.
Answer B is wrong because a large variation in torque values is a good indicator of a potentially warped head, hence cracks.
Answer C is wrong because finding small or otherwise invisible cracks can be greatly improved through the use of magnetic and dye type crack detectors.
Answer D is wrong because sanding residue from head surface areas improves the clarity of the surface without introducing undesirable scrapes and scratches. Smaller cracks become visible as they are filled with particles from sanding.

Question #18
Answer A is wrong because the head can be repaired by redressing the threads or installing a helicoil.
Answer B is correct because a helicoil is the standard fix for this condition.
Answer C is wrong because Technician A is wrong.
Answer D is wrong because Technician B is right.

Question #19
Answer A is correct because you use a straightedge and feeler gauge for detection of a transverse or longitudinal warp.
Answer B is wrong because this will not tell you if there is a warp on the flat side (fire deck) of the block. Differences in measurements could be caused by other deviations, and similar measurements exist when warps exist.
Answer C is wrong because only one technician is right.
Answer D is wrong because one technician is right.

Question #20
Answer A is wrong because discarded injectors are used to seal the ports at the injector sleeves.
Answer B is wrong because compressed air provides the medium for indicating leaks, because air will detect smaller leaks than liquids can escape through.
Answer C is wrong because threaded plugs are used to plug the return-line or crossover port at the other end of the head.
Answer D is correct because injection pumps are not used with unit injectors that have injector tubes.

Question #21
Answer A is correct because this is not a test that is performed by static measurement. It is a dynamic measurement.
Answer B is wrong because an uneven free height check of valve springs will reveal a weak spring preventing possible burnt valves.
Answer C is wrong because this test is performed to detect valve springs that might fail under load. If not performed the result could be a dropped valve.
Answer D is wrong because this inspection is performed to detect springs with slopped tops. If springs are not square, valves and guides may wear faster or stems may bind in guides.

Question #22
Answer A is wrong because missing seals would result in excessive oil consumption and the time period is too short for prolonged affects.
Answer B is correct because if the keepers are not seated properly the valve could drop in the time specified.
Answer C is wrong.
Answer D is wrong.

Question #23
Answer A is wrong because to allow the valve to cut through carbon deposits, a sharp edge needs to be formed by grinding the seat 1 degree steeper than the valve.
Answer B is wrong because rotators will remove carbon deposits from the seat and valve as the valve is rotated. Valves and seats are ground at the same angle when using a rotator.
Answer C is wrong because Technician A and B are both wrong.
Answer D is correct because both technicians are wrong. The one-degree difference is not necessary if valves are using rotators. The grinding wheel may be dressed with a one-degree steeper angle than the valve. The difference is used for a carbon breaking fit and better seal.

Question #24
Answer A is wrong because while cupping may appear quite often it is not normal. Cupping is usually a sign of overheating, excessive tappet clearance, overspeeding, or a weak valve spring. It may be obvious that other things should be corrected, but this valve needs replacement.
Answer B is wrong because it is expected that a new valve will meet specifications. Even after grinding to proper face angle, it is still necessary to check the valve margin. Any time you remove material from the valve it is important that adequate metal exists to prevent premature burning of the head.
Answer C is wrong because both technicians are wrong.
Answer D is correct because neither technician is right.

Question #25
Answer A is wrong because it is necessary to refer to the manufacturer's procedures for removing and installing sleeves because not all sleeves are the same and special tools may be required.
Answer B is wrong because this test is performed after the sleeves are installed to ensure the seal is good.
Answer C is wrong because this check is required to confirm that the injector position is correct. An incorrect height can cause improper cylinder operation by causing the injector spray pattern to strike the cylinder wall instead of the piston.
Answer D is correct because the replacement of an injector nozzle while possible is the LEAST likely needed repair.

Question #26
Answer A is correct because when placing pencil marks around the face of the valve about 1/8 in. (3 mm) apart and snapping the valve against its seat, if all but two lines are found to be visibly broken, the valve needs to be reground.
Answer B is wrong because the unbroken lines indicate that the valve seat is not concentric. This out-of-round condition will cause further damage to the valve and seats must be reground before installing the valve.
Answer C is wrong.
Answer D is wrong.

Question #27
Answer A is wrong because the inside micrometer is used for measuring inside measurement and comes with precision extensions that can be added to span great diameters. This is ideal for checking sleeves.
Answer B is wrong because a cylinder gauge is typically used to check the cylinder sleeve for out-of-roundness.
Answer C is correct because a dial indicator is not typically used to detect a defective sleeve (liner); it is used in the process to check liner height.
Answer D is wrong because a snap gauge or bore gauge are used to check the cylinder sleeve (liner) for out-of-roundness.

Question #28
Answer A is wrong because after wet-type cylinder sleeves have been removed, if the packing ring area and counter bore lip show signs of light rust and scale, you remove the rust.
Answer B is correct because one uses 100 to 120 grit emery paper to remove rust and scale and check for erosion and pitting.
Answer C is wrong.
Answer D is wrong.

Question #29
Answer A is correct because a piston skirt crack can be caused by improperly installed connecting rods with the connecting rod and bottom of the cylinder line damaged.
Answer B is wrong because cylinder liner to piston clearance will not cause skirt cracking.
Answer C is wrong.
Answer D is wrong.

Question #30
Answer A is correct because a heel-to-toe measurement can provide different readings, but they are still within specification due to excessive grooving of the cam lobe.
Answer B is wrong because the lobe shows signs of grooving and if it is more than 0.003 in. (0.076 mm), the cam needs to be replaced. If the timing of valves or the injector were to shift this much during operation it could cause unbalanced cylinder operation and stress on the crank as well as unstable operation of engine.
Answer C is wrong.
Answer D is wrong.

Question #31
Answer A is a good choice because you need to check the bearings and crankshaft before cleaning the crank. Yet, it is wrong because both technicians are right.
Answer B is also a good choice because you also check the crank after cleaning. Yet, it is wrong because both technicians are right.
Answer C is correct because both technicians are right.
Answer D is wrong because both technicians are right.

Question #32
Answer A is wrong because in order to sense engine speed and load changes the governor must have a direct link to the engine speed.
Answer B is wrong because to properly maintain the speed at a constant rate the governor must be able to directly affect the necessary changes in fuel to the injectors.
Answer C is wrong because to be variable the governor must be able to sense speed change requests from the operator.
Answer D is correct because the governor is dependent upon operator input and therefore not independent.

Question #33
Answer A is wrong because the fluid in the sensor expands as the temperature increases and is included in the older design.
Answer B is wrong because the fluid in the tube transmits the expanding fluid pressure of the sensor to the gauge in the mechanical gauge.
Answer C is correct because compensation resistors are used in later electronic sensors.
Answer D is wrong because the hydraulic meter movements respond to the pressure developed by the expanding fluids.

Question #34
Answer A is wrong because the gear pump is necessary to draw fuel from the pump.
Answer B is wrong because the throttle linkage provides a link to the accelerator to determine the mean operating speed.
Answer C is wrong because the pulsation damper is necessary to correct for pressure spikes caused by the gear pump as it draws fuel from the tank.
Answer D is correct because Cummins PT systems do not use transfer or lift pumps.

Question #35
Answer A is wrong because you cannot fill the feeler gauge into the cylinder with the piston in place. You have very limited space (only about .005 of an inch).
Answer B is correct because the piston ring end gap is checked with only the rings in the cylinder. Ring gap is an installed clearance with the ring in the cylinder and you can fit the piston and ring and measure the gap.
Answer C is wrong.
Answer D is wrong.

Question #36
Answer A is wrong because some bearings use a grove to direct oil around the shell not the notch.
Answer B is wrong because this is not a defect; the bearing shell is manufactured that way.
Answer C is wrong because oversized bearings have to be measured to determine their size.
Answer D is correct because this item called a tang keeps the bearing shell from turning.

Question #37
Answer A is correct because full load operation in excess of eight hours is permitted and therefore the exception.
Answer B is wrong because ignition of fluid in the vibration damper will result in a distortion in the shape of the vibration damper, resulting in excess vibration of the crankshaft and uneven wearing of bearing shells.
Answer C is wrong because teeth missing from the ring gear will cause an out of balance condition to exist, contributing to uneven bearing wear.
Answer D is wrong because loose torque converter bolts will result in excess vibration at the rear of the engine, and may cause the rear main to wear faster then the other bearings.

Question #38
Answer A is a good choice because you use a master bar to check the main bore alignment. Yet it is wrong because both technicians are right.
Answer B is a good choice because one uses an inside micrometer to check the main bore diameter for an out-of-round condition. Yet it is wrong because both technicians are right.
Answer C is correct because both technicians are right due to the above reasons.
Answer D is wrong because neither technician is wrong.

Question #39
Answer A is wrong because the passage would have to be clogged, not open to create this problem.
Answer B is wrong because incorrect oils will not cause this condition.
Answer C is correct because, depending on the design, an open in the wiring to the sensor will cause a no pressure reading.
Answer D is wrong because the wrong sensor will show the wrong reading, not the lack of any reading.

Question #40
Answer A is correct because regardless of the type of oil pump used, all mated surfaces must be checked for wear.
Answer B is wrong because most oil pumps receive limited wear and parts can be easily inspected and replaced if needed.
Answer C is wrong because Technician B is wrong.
Answer D is wrong because Technician A is right.

Question #41
Answer A is wrong because a visual inspection for cracks is a must.
Answer B is wrong because a visual inspection of gaskets surface will insure a tight seal.
Answer C is correct because magna fluxing of housing assembly and metal parts to detect small cracks is typically not performed.
Answer D is wrong because a visual inspection of all passageways is necessary to insure oil flows.

Question #42
Answer A is wrong because low oil pressure has no effect on cooler operation.
Answer B is wrong because even very high oil pressure would not cause a cooler failure.
Answer C is correct because cooler failures result in the mixing of oil and water.
Answer D is wrong because the engine oil would have water in it. When the pressure drops below 15 psi water will be forced through the oil cooler leak and deposit in the engine oil pan and result in higher oil level.

Question #43
Answer A is correct because the engine manufacturer used this key to offset the timing and it must be replaced with the same type.
Answer B is wrong because this is an offset key used by the manufacturer to offset the timing of the cam. The shift in timing may have been necessary for proper operation of some special option added to the standard engine or a special purpose application.
Answer C is wrong.
Answer D is wrong.

Question #44
Answer A is wrong because this is a generally accepted standard, but belts should be changed if they have any cracks in them. They can fail at any time and cause the pump to stop.
Answer B is correct because a drop of water coming from a hole in the bottom of the pump every five–ten minutes is the least likely cause of a complete cooling system failure.
Answer C is wrong because a loose clamp will cause a weak spot in the cooling system. Failure to correct this could cause the hose to blow under heavy loading conditions.
Answer D is wrong because if rust or scum is visible then it could already be clogging internal passages in the radiator or engine.

Question #45
Answer A is wrong because back flushing the radiator in the vehicle has limited success and can even clog passages or cause a rupture.
Answer B is wrong because the system will bleed off trapped air and is the preferred method when refilling a cooling system.
Answer C is wrong because both technicians are wrong.
Answer D is correct because both technicians are wrong.

Question #46
Answer A is wrong because SCA stands for supplemental coolant additive which is a corrosion inhibitor additive. Adding more antifreeze does not change an SCA reading.
Answer B is wrong because continuing to run the truck until the next PMI will not change the SCA reading.
Answer C is correct because one needs to drain the entire coolant system and add the proper SCA mixture to achieve the correct percentage of SCA.
Answer D is wrong because this will not change SCA readings.

Question #47
Answer A is a good choice because the in-line pump pressurizes, meters, times, and delivers fuel to injection nozzles. Yet it is wrong because both technicians are right.
Answer B is also a good choice because the injector lines between the pump and injection must be made of steel to handle high pressure fuel. Yet it is wrong because both technicians are right.
Answer C is correct because both technicians are right.
Answer D is wrong because both technicians are right.

Question #48
Answer A is correct because if the cam follower is worn the injector will be mistimed and have an insufficient stroke resulting in low fueling and a misfire.
Answer B is wrong because a restricted fuel filter will cause fuel starvation and low power for the entire engine, not just one cylinder.
Answer C is wrong because high fuel pressure will not affect injector operation in a rocker arm activated unit injector system like Detroit Diesel.
Answer D is wrong because an injector tube (sleeve) failure will not affect a misfire.

Question #49
Answer A is correct because these lines must be replaced as a set.
Answer B is wrong because these are high pressure lines and removal of a single one will require bending the line causing damage to the system.
Answer C is wrong.
Answer D is wrong.

Question #50
Answer A is wrong because a faulty injector may cause the engine to run rough, but should still start.
Answer B is wrong because a defective starter cutout relay would cause the starter to stay engaged after the engine has started.
Answer C is correct because if the energized to run (ETR) or engine shut off (ESO) solenoid fails no fuel will flow to the metering valve and pumping plungers resulting in a no-start.
Answer D is wrong because a leaky intake hose will cause possible future damage to turbocharger and valves but the engine should still start.

Question #51
Answer A is correct because worn turbo turbine blades are an indication of a loose or damaged intake hose between the filter and the turbo.
Answer B is wrong because the aftercooler cools output air not the inlet air.
Answer C is wrong because only one technician is right.
Answer D is wrong because only one technician is right.

Question #52
Answer A is correct because the air shutdown valve on a blower is used as an emergency shutdown device.
Answer B is wrong because this type of regular air shutdown can shock the seals out of the blower. Termination of fuel delivery is the normal shutdown process on a diesel engine.
Answer C is wrong because only one technician is right.
Answer D is wrong because only one technician is right.

Question #53
Answer A is correct because most diesels use either a regulating valve or tube that goes from the crankcase via the rocker cover to the air cleaner or intake manifold. Engine negative pressure (vacuum) acts through this valve or tube to depressurize the crankcase and overcome the effects of high pressure gases due to high compression. If this line, valve, or air filter becomes restricted, the crankcase is no longer depressed and the increased pressure causes leaks and higher blowby gases.
Answer B is wrong because the difference in pressure that one reads in the crankcase is a very small amount and therefore measured in inches of water. One inch of mercury equal 13.7 inches of water.
Answer C is wrong because only one technician is right.
Answer D is wrong because only one technician is right.

Question #54
Answer A is a good choice because an intercooler failure could cause the egress of water (coolant) into one of the cylinders. Yet it is wrong because both technicians are right.
Answer B is also a good choice because a hairline crack might not have been noticed in the previous repair without a dye check and cold cause water in a cylinder.
Answer C is correct because both technicians are right.
Answer D is wrong because both technicians are right.

Question #55
Answer A is wrong because a dry oil bath air filter would not account for high end overboost. This is a condition the wastegate corrects.
Answer B is correct because if the wastegate malfunctions too much boost can occur.
Answer C is wrong because only one technician is right.
Answer D is wrong because only one technician is right.

Question #56
Answer A is a good choice because in the Jacobs brake shown, the master piston at point A in the figure contains seals or O-rings that must be inspected and replaced. Yet A is not the right answer because both technicians are right.
Answer B is also a good choice because for the exhaust brake to work properly, the adjustment at the slave piston point B must open the exhaust valves before injection occurs. Yet B is not the right answer because both technicians are right.
Answer C is correct because both technicians are right.
Answer D is wrong because both technicians are right.

Question #57
Answer A is correct because one of the switches at C must be located at the fuel pump or accelerator pedal to detect when the accelerator is pressed.
Answer B is wrong because the Jacobs system will not affect the engine operation if any open occurs between the battery and the coil.
Answer C is wrong because only one technician is right.
Answer D is wrong because only one technician is right.

Question #58
Answer A is wrong because you do use a manometer to check the amount of vacuum or pressure.
Answer B is wrong because you do remove the air filter to calibrate the manometer.
Answer C is wrong because you do test with the air filter installed for restriction.
Answer D is correct because placing a piece of cardboard over the air intake to calibrate the manometer is an erroneous process.

Question #59
Answer A is a good choice because gray or black smoke could be an indication of low blower air box pressure. Yet this is the wrong answer because both technicians are right.
Answer B is also a good choice because low blower air box pressure could be caused by leaky end plate gaskets. Yet this is the wrong answer because both technicians are right.
Answer C is correct because both technicians are right.
Answer D is wrong because both technicians are right.

Question #60
Answer A is a good choice because to remove the surface charge from a recently charged battery, you apply a 300-ampere load across the adapters for 15 seconds. Yet this is the wrong answer because both technicians are right.
Answer B is a good choice because you do not remove the surface charge from batteries which have been in storage. Yet this is the wrong answer because both technicians are right.
Answer C is correct because both technicians are right.
Answer D is wrong because both technicians are right.

Question #61
Answer A is wrong because excessive exhaust back pressure will reduce airflow through engine and prevent good scavenging and reduce power out.
Answer B is wrong because of ineffective scavenging the cylinder temperature will increase.
Answer C is correct because poor combustion will not result from excess exhaust back pressure. Therefore it is the exception.
Answer D is wrong because excessive exhaust back pressure will reduce airflow through engine and prevent good savaging and reduce power out.

Question #62
Answer A is correct because a loss of engine lubrication oil through the breather tube is an indication of excess crankcase pressure.
Answer B is wrong because a slight crankcase pressure is desirable.
Answer C is wrong because only one technician is right.
Answer D is wrong because only one technician is right.

Question #63
Answer A is correct because PID stands for parameter identifier.
Answer B is wrong.
Answer C is wrong.
Answer D is wrong.

Question #64
Answer A is a good choice because when removing low pressure lines, inspect for damage to lines and connections. Yet this is wrong because both technicians are right.
Answer B is also a good choice because you need to inspect seals and O-rings for serviceability and replace as necessary. Yet this is wrong because both technicians are right.
Answer C is correct because both technicians are right.
Answer D is wrong because both technicians are right.

Question #65
Answer A is correct because if the snap pressure is low it may be necessary to change the throttle restriction.
Answer B is wrong because AFC pumps will not be accurate because they receive corrections from the manifold vacuum.
Answer C is wrong because if the throttle travel is incorrect ensure it is in the full open, not closed.
Answer D is wrong because if there is an incorrect low idle just change the spring shims.

Question #66
Answer A is wrong because on some OEMs, DTCs can also be cleared by removal of a battery cable.
Answer B is correct because on some trucks DTCs can be cleared by removing a battery cable.
Answer C is wrong because only one technician is right.
Answer D is wrong because only one technician is right.

Question #67
Answer A is a good choice because most generic scan tools by SAE J1978 standard can access any OBD II vehicle MIL DTCs. Yet it is wrong because both technicians are right.
Answer B is also a good choice because vehicle manufacturers have to provide MIL DTC commands to be OBD II certified.
Answer C is correct because both technicians are right.
Answer D is wrong because both technicians are right.

Question #68
Answer A is wrong because the positive lead to the relay must be removed.
Answer B is correct because you do not need to remove the negative lead to the relay.
Answer C is wrong because this is the device and they will be connected to the starter or a fender well.
Answer D is wrong because the ground cable to the battery must be disconnected to prevent a short circuit.

Question #69
Answer A is a good choice because you do not charge a sealed battery if the hydrometer is clear or yellow. Yet it is wrong because both technicians are right.
Answer B is also a good choice because the battery is charged when the built-in hydrometer dot is green. Yet it is wrong because both technicians are right.
Answer C is correct 184

Answer D is wrong because both technicians are right.

Question #70
Answer A is a good choice because a radiator pressure cap with a 9 stamped on it should open to release pressure at between 8 psi (55 kPa) to 10 psi (69 kPa). Yet it is wrong because both technicians are right.
Answer B is also a good choice because the same cap should open at about 0.6 psi (4.3 kPa) differential (vacuum) pressure. Yet it is wrong because both technicians are right.
Answer C is correct because both technicians are right.
Answer D is wrong because both technicians are right.

Question #71
Answer A is a good choice because when purging (bleeding) air from the fuel system it is necessary to pressurize the system with the hand primer. Yet it is wrong because both technicians are right.
Answer B is a good choice because purging (bleeding) air from the fuel system should be done any time fuel filters are changed. Yet it is wrong because both technicians are right.
Answer C is correct because both technicians are right.
Answer D is wrong because both technicians are right.

Question #72
Answer A is wrong because a technician should only connect a voltmeter in parallel with or across the circuit tested.
Answer B is wrong because positive terminal to ground only tests that circuit.
Answer C is wrong because a technician should only connect a voltmeter in parallel to the circuit being tested.
Answer D is correct because a technician should only connect a voltmeter in parallel with or across the circuit tested.

Question #73
Answer A is wrong because the figure indicates a swirling action to the air. This allows particles to be collected in the bottom "dust collector." It is not an oil bath air cleaner.
Answer B is correct because the bottom canister (labeled A) is a dust collector and must be emptied when servicing the engine.
Answer C is wrong because only one technician is right.
Answer D is wrong because only one technician is right.

Question #74
Answer A is wrong because the shutter drive piston is used to close the shutter which is working.
Answer B is correct because the shutter assembly springs open the shutters when the air is exhausted. Air pressure closes the shutters against spring pressure and these shutter springs open the shutters.
Answer C is wrong because this is used to close the shutter which is working.
Answer D is wrong because this is not used on an air system.

Question #75
Answer A is correct because you replace the fan if the blades are cracked.
Answer B is wrong because straightening badly bent blades may develop into fractures and cracks.
Answer C is wrong because only one technician is right.
Answer D is wrong because only one technician is right.

Question #76
Answer A is wrong because oil temperature is used to determine if the engine is warm or not.
Answer B is wrong because barometric pressure is used to determine outside condition of intake air.
Answer C is wrong because the PCM uses sensor inputs to vary the operation of glow plugs and indicator light.
Answer D is correct because the exhaust gas regulator (EGR) valve has no function in the glow plug system, so it is the exception. EGR actually lowers combustion temperature.

Question #77
Answer A is wrong because capsules are punctured when first installed.
Answer B is wrong because inlet check valve only opens during a vacuum condition and the outlet check valve only opens under a pressure condition.
Answer C is wrong because the plunger is used to draw fluid through one check valve and force it back out through the other check valve.
Answer D is correct because a solenoid valve is not used to release starting fluid.

Question #78
Answer A is a good choice because valve crossheads should be magna fluxed to check for cracks. Yet it is wrong because both technicians are right.
Answer B is also a good choice because the valve crossheads should be checked for out-of-roundness and excessive diameter. Yet it is wrong because both technicians are right.
Answer C is correct because both technicians are right.
Answer D is wrong because neither technician is right.

Question #79
Answer A is wrong because a unit injector that contains a seized needle valve will increase fuel pressure buildup when operating a priming pump. This would also cause the engine not to start but not cause the priming pump's failure to build pressure.
Answer B is correct because a leaking fuel filter housing would cause air to enter the system and pressure to bleed off.
Answer C is wrong because a restriction in these lines would cause primary pump vacuum and pressure to rise.
Answer D is wrong because a restriction in these lines would cause priming pump pressure to rise.

Question #80
Answer A is wrong because a ball gauge can be used.
Answer B is wrong because a snap gauge can be used.
Answer C is wrong because a dial gauge can be used.
Answer D is correct because an outside micrometer will not fit inside the guide bore or bear against it to measure movement.

Question #81
Answer A is wrong because constant use will cause excess strain on blower seals or piston oil rings due to high vacuum pressures.
Answer B is correct because the emergency air shutdown valves should not be used on a regular basis due to the strain created on the blower seals.
Answer C is wrong because only one technician is right.
Answer D is wrong because only one technician is right.

Question #82
Answer A is correct because cleaning and blowing out the grooves in the rocker cover with compressed air is an accepted process.
Answer B is wrong because oil will cause the silicone gasket to slip during installation. This could cause a bad seal.
Answer C is wrong because only one technician is right.
Answer D is wrong because only one technician is right.

Question #83
Answer A is a good choice because when you adjust a fuel shutoff solenoid, the solenoid plunger must be bottomed out and checked to meet manufacturer's specifications.
Answer B is a good choice because you adjust the external stop screw to the point where it touches the rack control.
Answer C is correct because both technicians are right.
Answer D is wrong because both technicians are right.

Question #84
Answer A is a good choice because to properly test an engine block for cracks, it should be submerged in a tank of water heated to between 180–200°F (82–93°C). Yet, A is wrong because both technicians are right.
Answer B is also a good choice because an engine block can be checked for water leaks by pressurizing it for about two hours without loss of pressure. Yet, B is wrong because both technicians are right.
Answer C is correct because both technicians are right.
Answer D is wrong because both technicians are right.

Question #85
Answer A is correct because adapter plates must fit snugly in sleeves to prevent cocking in the block.
Answer B is wrong because if the sleeve puller plate is too large it will cause damage to the block casting.
Answer C is wrong because only one technician is right.
Answer D is wrong because only one technician is right.

Question #86
Answer A is wrong because a fault can occur at a one to two ohms.
Answer B is wrong because a standard ohmmeter internal battery could cause damage to the module.
Answer C is wrong because both technicians are wrong.
Answer D is correct because both technicians are wrong.

Question #87
Answer A is correct because any generic scan tool can be used to make approved changes in a module, if the manufacturer's manual listed codes are used.
Answer B is wrong because a generic scan tool must be able to access approved change options for configuring modules.
Answer C is wrong because only one technician is right.
Answer D is wrong because only one technician is right.

Question #88
Answer A is a good choice because you use a straightedge across the exhaust manifold surfaces to align separate in-line heads on the engine block. Yet, A is wrong because both technicians are right.
Answer B is a good choice because you blow out head bolt holds to ensure bolts do not bottom out prematurely. Yet, B is wrong because both technicians are right.
Answer C is correct because both technicians are right.
Answer D is wrong because both technicians are right.

Question #89
Answer A is wrong because the stepper motor is not adjustable on a DS pulse width variable fuel solenoid style pump built by Stanadyne Diesel Systems. The PCM moves the stepper motor based on sensor input.
Answer B is correct because the GM Tech 2 scan tool is required to adjust the base timing.
Answer C is wrong because only one technician is right.
Answer D is wrong because only one technician is right.

Question #90
Answer A is a good choice because you never leave this tool in one position for too long. Yet, A is wrong because both technicians are right.
Answer B is a good choice because when this tool strikes a high spot in the cylinder, there should be an increase in drag on the drill. Yet, B is wrong because both technicians are right.
Answer C is correct because both technicians are right.
Answer D is wrong because both technicians are right.

Question #91
Answer A is correct because the use of improper tools can bend pins.
Answer B is wrong because pins cannot be straightened, the connector must be replaced again.
Answer C is wrong because improper or standard connectors may produce higher resistance.
Answer D is correct because grounding some circuits will cause damage to the PCM.

Question #92
Answer A is a good choice because gaskets used on internal solenoids must be replaced with solenoids. Yet, A is wrong because both technicians are right.
Answer B is a good choice because care should be taken to note positioning of insulating washers. Yet, B is wrong because both technicians are right.
Answer C is correct because both technicians are right.
Answer D is wrong because both technicians are right.

Question #93
Answer A is a good choice because bent push rods and tubes should be replaced, not straightened. Yet, A is wrong because both technicians are right.
Answer B is a good choice because push rods or tubes must be checked for proper fit to cam followers. Yet, A is wrong because both technicians are right.
Answer C is correct because both technicians are right.
Answer D is wrong because both technicians are right.

Question #94
Answer A is correct because you wait at least 10 minutes for codes to be issued from the scan tool.
Answer B is wrong because only a portion of the test must be completed within 10-minute scan tool reset time.
Answer C is wrong because this test is performed with the engine stopped.
Answer D is wrong because the scan tool will give an audible signal when setting is low or high.

Question #95
Answer A is a good choice because when the turbocharger lines and hoses become frayed or nicked they must be replaced, not repaired. Yet, A is wrong because both technicians are right.
Answer B is a good choice because turbocharger oil tubes must be checked for obstructions. Yet, B is wrong because both technicians are right.
Answer C is correct because both technicians are right.
Answer D is wrong because both technicians are right.

Question #96
Answer A is correct because this is the typical recommended antifreeze by all OEMs.
Answer B is wrong because Detroit Diesel recommends only ethylene glycol.
Answer C is wrong because Detroit Diesel recommends only ethylene glycol.
Answer D is wrong because Detroit Diesel recommends only ethylene glycol.

Question #97
Answer A is a good choice because the hydraulic lifter body should be inspected for scuffing and scoring. Yet, A is wrong because both technicians are right.
Answer B is a good choice because if the hydraulic lifter body is damaged, inspect the mating bore. Yet, B is wrong because both technicians are right.
Answer C is correct because both technicians are right.
Answer D is wrong because both technicians are right.

Question #98
Answer A is wrong because all identical belts should be changed.
Answer B is wrong because starting the belt on the edge of the pulley and cranking the engine will cause damage to belts.
Answer C is wrong because both technicians are wrong.
Answer D is correct because both technicians are wrong.

Question #99
Answer A is a good choice because air intake suppressors are used on normally aspirated engines that use a Jake brake. Yet, A is wrong because both technicians are right.
Answer B is a good choice because air intake suppressors are usually installed close to engine in the air intake system. Yet, B is wrong because both technicians are right.
Answer C is correct because both technicians are right.
Answer D is wrong because both technicians are right.

Question #100
Answer A is wrong because plugged nozzles will cause improper spray patterns.
Answer B is wrong because a stuck injection nozzle will cause injectors to dribble.
Answer C is correct because improper timing of injectors will not affect spray pattern and is the least likely cause of this condition.
Answer D is wrong because excessive injector-to-follower clearance will prevent adequate fuel pressurization for injection.

Question #101
Answer A is wrong because a short circuit causes high current flow, possibly damaging the meter. Generally, a short finder is used along with an ohmmeter.
Answer B is correct because if the battery is at less than the specified voltage, the current measured will be incorrect.
Answer C is wrong because only one technician is right.
Answer D is wrong because only one technician is right.

Question #102
Answer A is a good choice because petroleum jelly can prevent cable end corrosion but is the wrong choice because both A and B are both correct.
Answer B is a good choice because protective pads and grease will also reduce cable end corrosion but since both Technicians A and B are both correct, B is wrong.
Answer C is correct because both technicians are right.
Answer D is wrong because both technicians are right.

Question #103
Answer A is wrong because excessive smoke could be caused by a problem unrelated to the internal engine parts.
Answer B is correct because by discussing the problem with the owner or operator a correct diagnosis may occur. This is part of a diagnostic strategy.
Answer C is wrong.
Answer D is wrong.

Question #104
Answer A is wrong because cams are replaced if they do not meet visual inspection or specifications.
Answer B is wrong because cams are replaced if they do not meet visual inspection or specifications.
Answer C is wrong.
Answer D is correct.

Answers to the Test Questions for the Additional Test Questions Section 6

1. A
2. C
3. B
4. C
5. A
6. B
7. C
8. B
9. C
10. C
11. B
12. B
13. D
14. C
15. B
16. A
17. C
18. D
19. C
20 B
21. D
22. B
23. B
24. D
25. B
26. A
27. D
28. C
29. A
30. A
31. C
32. C
33. C
34. C
35. A
36. C
37. D
38. C
39. B
40. B
41. C
42. B
43. D
44. A
45. D
46. A
47. C
48. B
49. C
50. D
51. D
52. C
53. C
54. A
55. A
56. A
57. D
58. C
59. C
60. C
61. C
62. C
63. C
64. D
65. A
66. A
67. B
68. B
69. C
70. D
71. C
72. C
73. D
74. C
75. C
76. C
77. D
78. A
79. D
80. A
81. A
82. C
83. C
84. C
85. B
86. C
87. C
88. A
89. C
90. B
91. B
92. B
93. A
94. C
95. C
96. A
97. C
98. C
99. B
100. C
101. D
102. C
103. C
104. A
105. A
106. B
107. C
108. A
109. C
110. D
111. C
112. D
113. B
114. C
115. A
116. B
117. D
118. A
119. C
120. C
121. B
122. A
123. D
124. A
125. C
126. A
127. B
128. D
129. C
130. B
131. A
132. D
133. C
134. A
135. C
136. B
137. D
138. A
139. C
140. A
141. C
142. B
143. D
144. A
145. A
146. B
147. C
148. D

149. C
150. B
151. D
152. C
153. C
154. A
155. B
156. B
157. D
158. A
159. B
160. A
161. D
162. A
163. B
164. D
165. D
166. C
167. B
168. D
169. B
170. C
171. B
172. A
173. C
174. A
175. C
176. A
177. D
178. B
179. C
180. B
181. D
182. A
183. C
184. B
185. A
186. D
187. C
188. B
189. C
190. B
191. C
192. B
193. A
194. C
195. A
196. B
197. C
198. D

Explanations to the Answers for the Additional Test Questions Section 6

Question #1
Answer A is correct because this is an easy check to make and the only thing wrong may be a loose head.
Answer B is wrong because for best results crack detectors should be used only after the head has been removed and cleaned.
Answer C is wrong because the head bolts should have been checked before removing the head.
Answer D is wrong because a check of the head bolt torque should be accomplished before removing the head to remove deposits or perform a pressure test.

Question #2
Answer A is a good choice because the pressure raises the boiling point of the coolant and if the boiling point is lower the engine may overheat. Yet A is wrong because both technicians are right.
Answer B is a good choice because the cap is designed to release pressure above its rating. Yet B is wrong because both technicians are right.
Answer C is correct.
Answer D is wrong.

Question #3
Answer A is wrong because when the ball of the ball peen hammer is lightly tapped on the seat and head adjacent to the seat the sounds are compared. Similar ringing sound indicate a tight fitting seat, a dull sound or movement indicates a loose fitting seat.
Answer B is correct because sharpening a chisel to remove defective seats is not an approved procedure and therefore the least likely one to use.
Answer C is wrong because after grinding the seats ensure the opening is round using a concentricity indicator. If the seat is not round then it will leak.
Answer D is wrong because the valve seat grinding stone needs to be dressed for the specific valve seat angle that is being used.

Question #4
Answer A is a good choice because you cannot turn the camshaft bearing surface typically due to insufficient thickness or the cam is made of nodular cast iron and should not be resurfaced. Yet this choice is wrong because both technicians are right.
Answer B is also a good choice because a good camshaft inspection requires you to inspect the cam gear. Yet this choice is wrong because both technicians are right.
Answer C is correct because both technicians are right.
Answer D wrong because both technicians are wrong..

Question #5
Answer A is correct because a straightedge and feeler gauge is the right combination.
Answer B is wrong because the dial indicator and step block may provide a possibility of obtaining the measurement. The caliper and ruler are unnecessary.
Answer C is wrong because a micrometer is used for taking outside measurements and the point gauge is used to check orifices on the injector. Neither will provide the necessary type measurement.
Answer D is wrong because these tools will not provide measurements to the tolerances required for a valve head height check.

Question #6
Answer A is wrong because the pump shown is the Rexroth high pressure axial engine oil pump that provides 43 to 3,000 psi to operate the electronic hydraulic unit injectors (HEUI). Diesel fuel is supplied by a transfer pump to the common rail at low pressure regulated at 29 psi (200 kPa).
Answer B is correct because this fluid is coming from the engine oil sump. This lubricating oil pressure (almost 3,000 psi) is used by the injectors to produce a fuel injection pressure of almost 21,000 psi. If this pump produces lower than its normal pressure range it will affect the injection rate and therefore the power.
Answer C is wrong.
Answer D is wrong.

Question #7
Answer A is wrong because under run is not an affect of governor instability.
Answer B is wrong because high idle is the affect of a high idle setting.
Answer C is correct because hunting is the result of an incorrect governor buffer setting.
Answer D is wrong because low idle is the affect of a low idle setting.

Question #8
Answer A is wrong because this setup may cause damage to the charging system if terminals become shorted.
Answer B is correct because one makes the connections first so a short will not damage the good vehicle's charging system.
Answer C is wrong.
Answer D is wrong.

Question #9
Answer A is a good choice because the cause may be found by carefully inspecting the vibration damper. Yet A is wrong because both technicians are right.
Answer B is a good choice because the flywheel or torque converter needs to be closely inspected as a possible cause. Yet B is wrong because both technicians are right.
Answer C is correct.
Answer D is wrong.

Question #10
Answer A is a good choice because to check exhaust backpressure you connect a manometer to the exhaust manifold and check against manufacturer's specifications. Yet A is wrong because both technicians are right.
Answer B is a good choice because if a plug has not been provided for this test, drill and tap the exhaust manifold pipe. Yet B is wrong because both technicians are right.
Answer C is correct.
Answer D is wrong.

Question #11
Answer A is wrong because both use governor control.
Answer B is correct because this device improves driveability.
Answer C is wrong because both of these pumps use mechanical governors.
Answer D is wrong because fuel pressurization is accomplished in the injector; a duel action pump is not needed.

Question #12
Answer A is wrong because excess unburned fuel that has not reached combustion temperature will produce white smoke.
Answer B is correct because faulty or improperly installed rings may result in the engine producing a white smoke.
Answer C is wrong because this will cause burning of oil or water that causes blue smoke and white smoke.
Answer D is wrong because this condition will cause low operating temperature and failing of the injectors that produce black smoke.

Question #13
Answer A is wrong because you time in-line pumps with a timing pin not distributor pumps.
Answer B is wrong because the spill method is used on in-line pumps.
Answer C is wrong because the rack adjustment has no effect on timing in a distributor system.
Answer D is correct because a distributor pump is timed similar to an ignition distributor. Rotating the pump changes the start of injection in relation to crank position.

Question #14
Answer A is a good choice because if the air restriction is too excessive the filter or intake tubes need to be inspected. Yet A is wrong because both technicians are right.
Answer B is a good choice because the engine speed should be the same for both readings. Yet B is wrong because both technicians are right.
Answer C is correct.
Answer D is wrong.

Question #15
Answer A is wrong because problems may occur at a high rated speed load.
Answer B is correct because the engine must be tested across a range of operating speeds.
Answer C is wrong.
Answer D is wrong.

Question #16
Answer A is correct because coolant may form a pool of green or blue liquid below the component that is leaking.
Answer B is wrong because not all leaks are external. Sometimes, like a defective head gasket, the engine may consume fluids or expel them out as part of the exhaust. Monitoring of fluid levels and engine performance may be the only way to detect most internal leaks.
Answer C is wrong.
Answer D is wrong.

Question #17
Answer A is a good choice because pistons should be laid out in the order in which they came out. Yet A is wrong because both technicians are right.
Answer B is a good choice because installing an improper injector nozzle could cause scoring and cracking of pistons skirts. Yet B is wrong because both technicians are right.
Answer C is correct.
Answer D is wrong.

Question #18
Answer A is wrong because the end of the valve stem is ground flat to ensure that the tappet clearance remains the same even if the valve rotates.
Answer B is wrong because any time the valve is to be reused it must be inspected and reground to ensure proper sealing of the valves. Failure to re-face valves can result in a premature overhaul of the head.
Answer C is wrong.
Answer D is correct.

Question #19
Answer A is a good choice because on a PT system you back out the torque screw and high idle screws must be backed out about 5 turns. Yet A is wrong because both technicians are right.
Answer B is a good choice because the pump must rotate to produce 1,000 strokes to collect the proper amount of fuel. Yet B is wrong because both technicians are right.
Answer C is correct.
Answer D is wrong because neither technician is wrong.

Question #20
Answer A is wrong because battery posts are easily replaced.
Answer B is correct because you should only replace the battery when it fails the appropriate tests.
Answer C is wrong because battery cables are easily replaced.
Answer D is wrong because battery clamps are easily replaced.

Question #21
Answer A is wrong because the oil pump by-pass valves are inspected and adjusted any time the oil pump is removed.
Answer B is wrong because only items that fail inspection are replaced unless they are a warranty item.
Answer C is wrong.
Answer D is correct.

Question #22
Answer A is wrong because the engine must be off because the linkages are removed.
Answer B is correct because the key must be on and the engine off.
Answer C is wrong because the key must be on to power the diagnostic.
Answer D is wrong because the diagnostic tool can not start the engine.

Question #23
Answer A is wrong because the fuel solenoid pushes the governor to the no fuel position.
Answer B is correct because you push the governor linkage to the no fuel position.
Answer C is wrong because the fuel solenoid pushes the governor to the no fuel position.
Answer D is wrong because the fuel solenoid pushes the governor to the no fuel position.

Question #24
Answer A is wrong because this is a required task when installing a starter.
Answer B is wrong because this is a required task when installing a starter.
Answer C is wrong because this is a required task when installing a starter.
Answer D is correct because starter bushings are located inside a starter.

Question #25
Answer A is wrong because the inline injector pump is timed by an internal pump cam and plunger, not the engine cam and injector.
Answer B is correct because timing at the injector pump is accomplished by a camshaft inside the pump.
Answer C is wrong.
Answer D is wrong.

Question #26
Answer A is correct because a multimeter is not used to test battery capacity.
Answer B is wrong because an ammeter is used to test battery capacity.
Answer C is wrong because a voltmeter is used to test battery capacity.
Answer D is wrong because a carbon pile load is used to test battery capacity.

Question #27
Answer A is wrong because suction leaks in the fuel supply system will cause air to enter the fuel system.
Answer B is wrong because incorrect throttle travel will prevent the fuel rack from achieving correct fuel position for load.
Answer C is wrong because it will restrict fuel flow to injectors.
Answer D is correct because a leak in the injector return line will cause low fueling and low power.

Question #28
Answer A is a good choice because the lines shown are high-pressure fuel lines. Yet A is wrong because both technicians are right.
Answer B is a good choice because all fuel connections must be capped when they are disconnected. Yet B is wrong because both technicians are right.
Answer C is correct.
Answer D is wrong.

Question #29
Answer A is correct because excessive lubrication levels in the crankcase is the least likely cause.
Answer B is wrong because this will cause heat buildup and metal-to-metal contact, which causes scoring and cracks.
Answer C is wrong because this will cause uneven pressures and excessive heat causing the piston to become fatigued and fracture.
Answer D is wrong because this will cause the piston to become fatigued and fracture.

Question #30
Answer A is correct because the turbine outlet is connected to the exhaust system.
Answer B is wrong because the output of the compressor needs to be cooled so it is sent to the aftercooler.
Answer C is wrong because after the air is cooled it is applied to the cylinders through the intake manifold.
Answer D is wrong because the filtered air is applied to the inlet of the compressor.

Question #31
Answer A is a good choice because you replace springs that are fractured or broken, out of square, or do not meet the specifications. Yet A is wrong because both technicians are right.
Answer B is a good choice because companion springs under the same valve bridge as defective springs must also be replaced. Yet B is wrong because both technicians are right.
Answer C is correct.
Answer D is wrong.

Question #32
Answer A is a good choice because "lugging" is when the throttle is in full fuel position and cannot pick up speed under load. Yet A is wrong because both technicians are right.
Answer B is a good choice because more fuel is burned than needed in the lugged condition. Yet B is wrong because both technicians are right.
Answer C is correct.
Answer D is wrong.

Question #33
Answer A is a good choice because when removing low-pressure fuel lines ensure the ground cable is removed from the battery. Yet A is wrong because both technicians are right.
Answer B is a good choice because line connections must be capped to prevent dirt from entering the fuel system. Yet B is wrong because both technicians are right.
Answer C is correct.
Answer D is wrong.

Question #34
Answer A is a good choice because injector tip height must be checked to confirm proper clearance. Yet A is wrong because both technicians are right.
Answer B is a good choice because you perform a cylinder head leak test. Yet B is wrong because both technicians are right.
Answer C is correct.
Answer D is wrong.

Question #35
Answer A is correct because listening to the exhaust to determine which cylinder is misfiring is the least likely process.
Answer B is wrong because this is a likely process and unlike listing to the exhaust, this method will usually reveal which cylinder to conduct further test on.
Answer C is wrong because this procedure will not only point to the suspect cylinder but it will also confirm that the injector is not firing.
Answer D is wrong because this method is extremely useful in reducing possibilities when the sound seems to be coming from everywhere.

Question #36
Answer A is a good choice because the customer should not disconnect the battery before seeking a service center when an emission control problem is detected. Yet A is wrong because both technicians are right.
Answer B is a good choice because after clearing DTCs by disconnecting the battery, it may be necessary to operate the vehicle several times to resume normal operation. Yet B is wrong because both technicians are right.
Answer C is correct.
Answer D is wrong.

Question #37
Answer A is wrong because this acronym refers to the OBD II codes used to turn on instrument panel indicator lights.
Answer B is wrong because this acronym refers to the code assigned and stored for a specific vehicle fault.
Answer C is wrong because this acronym refers to the standards established for all new vehicles. These standards establish requirements for the detection and storage of faults while a vehicle is operating.
Answer D is correct because it is a bogus process.

Question #38
Answer A is a good choice because continual use of the air shutdown valve while the engine is under heavy loads or at high RPM will result in damage to blower oil seals. Yet A is wrong because both technicians are right.
Answer B is a good choice because before replacing lightly scored blower rotor blades try to dress them with a file. Yet B is wrong because both technicians are right.
Answer C is correct.
Answer D is wrong.

Question #39
Answer A is wrong because a clogged passage like a chinked line would result in slow response and faulty readings if the blockage is minimal, increasing until no reading is indicating.
Answer B is correct because the line has a leak. When an installed engine oil pressure gauge is compared to a test gauge at low engine rpm the installed gauge appears low, but at normal engine rpm, the installed gauge readings are correct.
Answer C is wrong.
Answer D is wrong.

Question #40
Answer A is wrong because the positive must be supplied to the relay.
Answer B is correct because installing negative lead to relay is the exception.
Answer C is wrong because these must be connected to operate the relay.
Answer D is wrong because this must be connected to provide a ground connection to all electrical circuit back to the battery.

Question #41
Answer A is a good choice because the battery term ampere hour refers to stored charge capacity of a battery. Yet A is wrong because both technicians are right.
Answer B is a good choice because a 75-ampere hour charge applied to a 200-ampere hour battery should turn the charge indicator green. Yet B is wrong because both technicians are right.
Answer C is correct.
Answer D is wrong.

Question #42
Answer A is wrong because this test is checking for excess resistance.
Answer B is correct because voltmeters measure voltage.
Answer C is wrong because this would damage the ohmmeter.
Answer D is wrong because this would damage the ammeter.

Question #43
Answer A is wrong because the linkage adjustments are used to calibrate closure.
Answer B is wrong because the ambient temperature affects thermostatic piston movement.
Answer C is wrong because the radiator temperature affects the thermostatic piston movement.
Answer D is correct. Air pressure is the exception because it is the medium used to operate the shutter, not regulate it.

Question #44
Answer A is correct because a connector found to be distorted from heat may cause problems with low-current signals.
Answer B is wrong because electrical connectors that have been heated may appear to provide good electrical connection when checked using direct current (DC) meters. Fractures in the wire may adversely affect the operation of data-flow at frequencies of greater than 10 kHz.
Answer C is wrong.
Answer D is wrong.

Question #45
Answer A is wrong because defective caps must be replaced.
Answer B is wrong because radiator tubes cannot stand high pressure.
Answer C is wrong because hot hoses are very flexible and will collapse when a vacuum is applied to the system. Constant collapsing of hoses will permanently distort the hose.
Answer D is correct because you cannot adjust or replace the spring in the cap.

Question #46
Answer A is correct because fan bearings should be submerged in cleaning solvent and blown out with air to remove dirt.
Answer B is wrong because they are not water cooled and any corrosion is reason to replace bearings.
Answer C is wrong.
Answer D is wrong.

Question #47
Answer A is a good choice because canister filters should be pre-filled to protect the bearings. Yet A is wrong because both technicians are right.
Answer B is a good choice because you need to remove the filter housing mounted by-pass valve and inspect when replacing the filter. Yet B is wrong because both technicians are right.
Answer C is correct.
Answer D is wrong.

Question #48
Answer A is wrong because an indication of only 1–10 minutes is used to confirm an operation system.
Answer B is correct because most glow plug systems will have the glow plug light on when the system has experienced a short circuit.
Answer C is wrong because the light does not relate to on and off time of glow plug on systems other than GM.
Answer D is wrong because the PCM or engine temperature sensor controls the glow plug system.

Question #49
Answer A is a good choice because end gaps on rings must be offset to prevent blowby. Yet A is wrong because both technicians are right.
Answer B is a good choice because when installing a piston and connection rod, studs should be covered to prevent damage to the crankshaft. Yet B is wrong because both technicians are right.
Answer C is correct.
Answer D is wrong.

Question #50
Answer A is wrong because the indication is water in the oil, this would allow coolant to enter the crankcase by way of oil passages in the head or block.
Answer B is wrong because the indication is water in the oil, this would allow coolant to enter the crankcase by way of oil passages in the head.
Answer C is wrong because the indication is water in the oil, this would allow coolant to mix with the oil in the oil cooler and return to the crankcase.
Answer D is correct because a worn oil pump is the least likely cause of this condition because of little or no contact with the engine oil.

Question #51
Answer A is wrong because check valves cannot be disassembled.
Answer B is wrong because plunger seals are replaceable.
Answer C is wrong.
Answer D is correct.

Question #52
Answer A is a good choice because valve rotators can be tested using a plastic mallet. Yet A is wrong because both technicians are right.
Answer B is a good choice because defective valve seals will cause oil consumption. Yet B is wrong because both technicians are right.
Answer C is correct.
Answer D is wrong.

Question #53
Answer A is a good choice because crosshead guide pins should be at right angles to the heads milled surface. Yet A is wrong because both technicians are right.
Answer B is a good choice because to check crosshead guide pin diameter with a micrometer, and compare to manufacturer's specifications is right. Yet B is wrong because both technicians are right.
Answer C is correct.
Answer D is wrong.

Question #54

Answer A is correct because the exception is that radiator and engine back-flushing is not recommended annually.

Answer B is wrong because you check coolant for corrosive elements even if the coolant is not going to be re-used.

Answer C is wrong because failure to bleed air from the coolant system will produce hot spots in the engine block.

Answer D is wrong because testing coolant for protection could prevent a major overhaul.

Question #55

Answer A is correct because when the manufacturer does not supply specifications, press the new guide in head to the same position the old guide was in.

Answer B is wrong because guide bores that are severely scored should be reamed out. Install over-sized valve guides.

Answer C is wrong.

Answer D is wrong.

Question #56

Answer A is correct because all gears in the timing geartrain must be inspected for tooth wear, including the crankshaft gear.

Answer B is wrong because this condition is not normal and the gears must be replaced; repair should not be considered.

Answer C is wrong.

Answer D is wrong.

Question #57

Answer A is wrong because this is good maintenance on any major component.

Answer B is wrong because these must be flat to make a good seal.

Answer C is wrong because these must be flat to make a good seal.

Answer D is correct because you must replace blower screens.

Question #58

Answer A is a good choice because before installing die cast rocker covers on a cylinder head, ensure the silicone gasket is secure in the cover groove to ensure a good seal. Yet A is wrong because both technicians are right.

Answer B is a good choice because you should always use new gaskets when installing valve covers. Yet B is wrong because both technicians are right.

Answer C is correct.

Answer D is wrong.

Question #59

Answer A is a good choice because manufacturers allow for movement in the intake system components by using durable rubber couplings, which are clamped at both ends. Yet A is wrong because both technicians are right.

Answer B is a good choice because if an air intake pipe, tube, or hose is loose or missing a clamp an inspection should be made of the turbocharger compressor and valves. Yet B is wrong because both technicians are right.

Answer C is correct.

Answer D is wrong.

Question #60
Answer A is a good choice because when testing a cylinder head for potential coolant leakage the coolant needs to be heated to operating temperature. Yet A is wrong because both technicians are right.
Answer B is a good choice because the coolant needs to be under pressure. Yet B is wrong because both technicians are right.
Answer C is correct.
Answer D is wrong.

Question #61
Answer A is a good choice because you remove the fuel shutoff solenoid disconnect wire from terminals, remove the ball adjuster from the control rack, and remove the screws holding the solenoid. Yet, A is wrong because both technicians are right.
Answer B is a good choice because when reinstalling, you position the solenoid and mount it using screws, then connect the ball adjuster to the rack and attach the wires only. Yet B is wrong because both technicians are right.
Answer C is correct.
Answer D is wrong.

Question #62
Answer A is a good choice because when pressure testing an engine block, the block need only be submerged in water for about 20–30 minutes and the water checked for bubbles. Yet A is wrong because both technicians are right.
Answer B is a good choice because cracked cylinder blocks should be replaced. Yet B is wrong because both technicians are right.
Answer C is correct.
Answer D is wrong.

Question #63
Answer A is a good choice because extreme caution is a must when working around operational turbochargers because the turbine turns between 60,000 to 100,000 rpm. Yet A is wrong because both technicians are right.
Answer B is a good choice because a compressor turbine wheel must be replaced if it has a nicked or cracked blade. Yet B is wrong because both technicians are right.
Answer C is correct.
Answer D is wrong.

Question #64
Answer A is wrong because this will cause excess exhaust back pressure preventing scavenging and aerating of the cylinder. Without a fresh supply of air in the cylinders you can not have combustion regardless of temperature.
Answer B is wrong because if the starter has a short in the armature, not enough torque can be developed to turn the engine over. An open would prevent the starter armature from developing a magnetic field necessary for rotation.
Answer C is wrong because a corroded battery terminal will limit necessary initial current flow to the starter to start the engine cranking. Most torque is developed when the starter first engages the flywheel. Once the starter has made contact the speed and torque drops and the current draw increases.
Answer D is correct because lack of ether is the LEAST likely cause of this no-start condition as described.

Question #65
Answer A is correct because you use a manometer to check pressure.
Answer B is wrong because failures may not occur at an idle. All operating speeds need to be checked.
Answer C is wrong.
Answer D is wrong.

Question #66
Answer A is correct because main bearing's shells should receive a light coat of oil before final installation.
Answer B is wrong because only one journal should be checked at a time unless others are readily available. Under no circumstance turn the crank while plastic gauge is installed.
Answer C is wrong.
Answer D is wrong.

Question #67
Answer A is wrong because it is a cylinder sleeve puller.
Answer B is correct because the tool as shown in the figure is used to bore out cylinder liners.
Answer C is wrong.
Answer D is wrong.

Question #68
Answer A is wrong because the cyclonic airflow indicated is for an airflow air filter not for lubrication.
Answer B is correct because this type air filter has a reduced efficiency at low rpm.
Answer C is wrong.
Answer D is wrong.

Question #69
Answer A is a good choice because before installation, head bolts must be cleaned and inspected for erosion or pitting. Yet A is wrong because both technicians are right.
Answer B is a good choice because the cylinder needs to be checked for foreign objects before installing the head. Yet B is wrong because both technicians are right.
Answer C is correct.
Answer D is wrong.

Question #70
Answer A is wrong because this type of primer has a glass or metal bowl to catch sediments.
Answer B is wrong because this type of primer has a plastic or metal strainer to block sediments before entering the pump and fuel lines.
Answer C is wrong because this type of primer has a hand operated plunger type pump.
Answer D is correct because the pop-off valve is the exception.

Question #71
Answer A is a good choice because when honing a cylinder liner the hone should shake strongly to provide a good cleaning. Yet A is wrong because both technicians are right.
Answer B is a good choice because once the honing has been started do not remove it, even to clean the stones. Yet B is wrong because both technicians are right.
Answer C is correct.
Answer D is wrong.

Question #72
Answer A is a good choice because with reduced power, you may need to bleed the air out of the intercooler. Yet A is wrong because both technicians are right.
Answer B is a good choice because when you have reduced power and high coolant temperature are noted on an engine with an intercooler, the air tubes in the cooler may be clogged. Yet B is wrong because both technicians are right.
Answer C is correct.
Answer D is wrong.

Question #73
Answer A is wrong because if rocker arm bushings are worn they must be replaced.
Answer B is wrong because if rocker arm shafts are worn they must be replaced.
Answer C is wrong because scored and pitted rocker arms may cause damage after hours of operation and should be replaced.
Answer D is correct because you do not tighten down the rocker arm tappet screws.

Question #74
Answer A is a good choice because small leaks can easily be detected because antifreeze does not evaporate and generally leaves a red or green colored buildup or stain nearby. Yet A is wrong because both technicians are right.
Answer B is a good choice because a defective thermostat is most frequently caused by repeated overheating of the engine. Yet B is wrong because both technicians are right.
Answer C is correct.
Answer D is wrong.

Question #75
Answer A is a good choice because a broken crank could be caused by the operator hot-riding the engine. Yet A is wrong because both technicians are right.
Answer B is a good choice because an out-of-round condition may exist on one or more of the main bores. Yet B is wrong because both technicians are right.
Answer C is correct.
Answer D is wrong.

Question #76
Answer A is a good choice because gauges work better than indicators because you can see when they are not working. Yet A is wrong because both technicians are right.
Answer B is a good choice because failure of a digital or electronic gauge may not be the fault of the actual gauge circuit. Yet B is wrong because both technicians are right.
Answer C is correct.
Answer D is wrong.

Question #77
Answer A is wrong because any generic scan tool must be able to access on board diagnostic and necessary user programmable selections.
Answer B is wrong because any generic scan tool must be capable of connecting to the data link connector.
Answer C is wrong.
Answer D is correct.

Question #78
Answer A is correct because you use a digital voltmeter to test grounds with power applied.
Answer B is wrong because an ohmmeter is used when power is not applied.
Answer C is wrong.
Answer D is wrong.

Question #79
Answer A is wrong because this measurement is used to calculate maximum resurfacing allocable on a head that is already warped. If warpage can be corrected by resurfacing, the head is used. The head is replaced if resurfacing. To remove the warpage would cause the deck-to-deck thickness to be out of tolerance.
Answer B is wrong because to properly determine valve guide height most manufacturers require special insertion tools.
Answer C is wrong because these types of measurements would prove inconclusive for warpage.
Answer D is correct because this measurement will determine the maximum allowable resurfacing.

Question #80
Answer A is correct because zero to five volts is the standard voltage reference value on non OBD II systems from WOT to the closed position.
Answer B is wrong because the voltage at full open is 0 and at full closed is 5.
Answer C is wrong because the voltage at full open is 0 and at full closed is 5.
Answer D is wrong because the voltage at full open is 0 and at full closed is 5.

Question #81
Answer A is correct because the micro connector is shown.
Answer B is wrong.
Answer C is wrong.
Answer D is wrong.

Question #82
Answer A is a good choice because a small leak can easily be detected because antifreeze does not evaporate and generally leaves a red or green colored build-up or stain nearby. Yet A is wrong because both technicians are right.
Answer B is a good choice because defective thermostat is most frequently caused by repeated overheating of the engine. Yet B is wrong because both technicians are right.
Answer C is correct.
Answer D is wrong.

Question #83
Answer A is a good choice because the internal shutdown solenoid in the figure must be tested after replacement. Yet A is wrong because both technicians are right.
Answer B is a good choice because to clean the old gasket off the cover and fuel pump when changing the shutdown solenoid is right. Yet B is wrong because both technicians are right.
Answer C is correct.
Answer D is wrong.

Question #84
Answer A is a good choice because to test the oil pressure gauge, it should mimic the instrument panel gauge in readings. Yet A is wrong because both technicians are right.
Answer B is a good choice because test readings should be taken during startup, varying operating ranges, and on shut down. Yet B is wrong because both technicians are right.
Answer C is correct.
Answer D is wrong.

Question #85
Answer A is wrong because this would reduce the scavenging and fresh air needed for combustion, thereby reducing the air to fuel mixture and power output.
Answer B is correct because high density air supplied to the cylinder is what is required and therefore the least likely cause of a power loss.
Answer C is wrong because restrictions in the airflow through the tubes will reduce the compression heat, thereby reducing the combustion temperature and power output. Restricted fins cause inefficient cooling, leading to hotter but less dense air supply to the cylinder. This less dense air reduces the compression heat reducing the efficiency of the combustion and power output.
Answer D is wrong because this would reduce the supply of necessary fuel at a power rating where maximum fuel supply is critical to injection and cooling of the injectors. There is a high probability that injectors will start failing, as well as fuel starvation and loss of power.

Question #86
Answer A is a good choice because when the throttle position sensor is adjusted on a fuel injection pump with electronic controls, on some applications, a gauge block is placed between the gauge boss and the maximum throttle travel screw. Yet A is wrong because both technicians are right.
Answer B is a good choice because the scan tool will signal when the setting is too low or too high. Yet B is wrong because both technicians are right.
Answer C is correct.
Answer D is wrong.

Question #87
Answer A is a good choice because ethylene glycol and methoxy propanol should not be mixed. Yet A is wrong because both technicians are right.
Answer B is a good choice because switching the type of antifreeze used, the engine must be flushed completely first. Yet B is wrong because both technicians are right.
Answer C is correct.
Answer D is wrong.

Question #88
Answer A is correct because oil in a hold-down bolt hole of a head can throw off the torque setting of that bolt.
Answer B is wrong because a bolt with stripped threads should be replaced.
Answer C is wrong.
Answer D is wrong.

Question #89
Answer A is a good choice because wet cylinder sleeves should be cleaned by immersion or with glass bead plating and then closely inspected. Yet A is wrong because both technicians are right.
Answer B is a good choice because a cylinder out-of-round condition can be detected by measuring at one point and rotating around the inside 180 degrees. Yet B is wrong because both technicians are right.
Answer C is correct.
Answer D is wrong.

Question #90
Answer A is wrong because the temperature is 50°F not 70°F and above.
Answer B is correct at 9.6V; See the chart.
Answer C is wrong because the temperature is 50°F not 30°F.
Answer D is wrong because the temperature is 50°F not 15°F.

Question #91
Answer A is wrong because welding may cause warping of the head surface of the block and weaken the metal between cylinders.
Answer B is correct because the block needs to be sent out to have the counterbore re-sleeved.
Answer C is wrong.
Answer D is wrong.

Question #92
Answer A is wrong because they should be convex.
Answer B is correct because if a lifter foot is pitted, the camshaft mating lobe should be inspected.
Answer C is wrong.
Answer D is wrong.

Question #93
Answer A is correct because a vibration occurrence rate that is equal to 1/2 the engine rpm might be caused by a bent cam, because it turns at 1/2 engine speed.
Answer B is wrong because this will create a vibration with the same occurrence rate as the engine speed.
Answer C is wrong.
Answer D is wrong.

Question #94
Answer A is a good choice because alignment between pulleys must not exceed 1/16 in. (1.59 mm) for each 12 in. (30.5 cm) of distance between pulley centers. Yet A is wrong because both technicians are right.
Answer B is a good choice because drive belts must not touch the bottom of grooves and protrude more than 3/32 in. (2.38 mm) above the top of the groove. Yet B is wrong because both technicians are right.
Answer C is correct.
Answer D is wrong.

Question #95
Answer A is a good choice because testing of the oil cooler is to be done between engine overhauls. Yet A is wrong because both technicians are right.
Answer B is a good choice because testing of the oil cooler is to be done during an engine overhaul. Yet B is wrong because both technicians are right.
Answer C is correct.
Answer D is wrong.

Question #96
Answer A is correct because an air intake suppressor can eliminate pulsation.
Answer B is wrong because wastegates are used to prevent over boost.
Answer C is wrong because an oil bath air cleaner will not stop the pulsation.
Answer D is wrong because variable cam timing will not stop the pulsation.

Question #97
Answer A is a good choice because PID codes are retrieved by using the diagnostic test set. Yet A is wrong because both technicians are right.
Answer B is a good choice because PID codes can be cleared by removing the negative battery cable. Yet B is wrong because both technicians are right.
Answer C is correct.
Answer D is wrong.

Question #98
Answer A is wrong because when battery potential is applied the diode is reversed biased.
Answer B is wrong because the applied voltage in this circuit is direct current.
Answer C is correct because the diode is used to shunt the stored potential when the coil is de-energized.
Answer D is wrong because this is a diode not an RFI filter.

Question #99
Answer A is wrong because the injector rocker arm is activating the valve not the other way around.
Answer B is correct because the master piston at position A senses the push rod upward motion to open exhaust valves.
Answer C is wrong because the slave piston senses the upward motion to open exhaust valves.
Answer D is wrong because the master piston does not operate the rocker arm, the rocker arm moves the device (valve).

Question #100
Answer A is a good choice because a PT injector system when that pressure of fuel against the angular surface of the injector needle exceeds the spring tension, the injector fires. Yet A is wrong because both technicians are right.
Answer B is a good choice because a unit injector system when the cam lobe presses the plunger of the injector down, pressurizes fuel and injects fuel into the cylinder. Yet B is wrong because both technicians are right.
Answer C is correct.
Answer D is wrong.

Question #101
Answer A is wrong because the valves are adjusted as the second item in the order of a Detroit Diesel tune-up.
Answer B is wrong because the low speed spring adjustment is the the end of the process.
Answer C is wrong because the governor gap is the fourth item in the process order.
Answer D is correct because the bridges must be balanced first before the valves or injectors are adjusted.

Question #102
Answer A is wrong because the prechamber is driven into the cylinder head with a plastic hammer and light pressure.
Answer B is wrong because there is no adjustment of the prechamber.
Answer C is correct because prechambers are fitted flush to the fire deck or plus .002 of an inch above it.
Answer D is wrong because there is no fit adjustment.

Question #103
Answer A is a good choice because on a computer-controlled diesel engine, corrosion on wire connectors can affect engine performance. Yet A is wrong because both technicians are right.
Answer B is a good choice because after repairing wire connectors they should be treated with a corrosion inhibitor. Yet B is wrong because both technicians are right.
Answer C is correct.
Answer D is wrong.

Question #104
Answer A is correct because a fan belt is the LEAST likely cause of an engine knock and does track with engine speed.
Answer B is wrong because camshaft noise will increase with engine rpm.
Answer C is wrong because crankshaft noise will increase with engine rpm.
Answer D is wrong because rocker arm noise will increase with engine rpm.

Question #105
Answer A is correct because over fueling can cause very black smoke to come out of the exhaust pipe. Fuel burned without air is smoke.
Answer B is wrong because the coolant level has no effect on an over fueling problem.
Answer C is wrong.
Answer D is wrong.

Question #106
Answer A is wrong because with the battery disconnected, you will not get any readings.
Answer B is correct because in the diagnostic strategy, you connect a scan tool or laptop to the system first.
Answer C is wrong because the injector lines have to be connected for testing.
Answer D is wrong because the key is turned on after the tool is hooked up.

Question #107
Answer A is wrong.
Answer B is wrong.
Answer C is correct because the tool shown in the figure is a hydrometer.
Answer D is wrong.

Question #108
Answer A is correct because a low voltage problem at the starter may be caused by a worn or shorted battery cable.
Answer B is wrong because a bad starter drive has no effect on voltage.
Answer C is wrong because only one technician is right.
Answer D is wrong because one technician is right.

Question #109
Answer A is wrong because the negative cable is removed first to prevent arching.
Answer B is wrong because the negative cable is removed first to prevent arching.
Answer C is correct because the negative cable is removed first.
Answer D is wrong because the negative cable is removed first to prevent arching.

Question #110
Answer A is wrong because injection pump internal components are lubricated by the diesel fuel.
Answer B is wrong because many diesel engines with line pumps lubricate the injection pump with oil.
Answer C is wrong because some engines use a port in the engine block.
Answer D is correct because the transfer pump typically does not provide pump lube, therefore it is the exception.

Question #111
Answer A is a good choice because starter fluid may be used. Yet, it is wrong because both technicians are right.
Answer B is a good choice because an immersion heater in the oil or coolant will help start the engine easier. Yet, it is wrong because both technicians are right.
Answer C is correct because both technicians are right.
Answer D is wrong because both technicians are right.

Question #112
Answer A is wrong because a plugged air filter will restrict airflow to the engine and cause a power loss.
Answer B is wrong because a fuel leak will cause a power loss.
Answer C is wrong because a plugged fuel filter will cause a power loss.
Answer D is correct because low coolant level will NOT cause a power loss.

Question #113
Answer A is wrong.
Answer B is correct because the figure shows a vibration damper.
Answer C is wrong.
Answer D is wrong.

Question #114
Answer A is a good choice because if the pressure in the radiator is too great, the pressure cap will release the excess pressure and fluid into an expansion tank. Yet A is wrong because both technicians are right.
Answer B is a good choice because most radiator pressure caps have a release pressure of 15 psi or lower. Yet B is wrong because both technicians are right.
Answer C is correct.
Answer D is wrong.

Question #115
Answer A is correct because a crack in the exhaust valve will NOT cause coolant in the engine oil, so it is the exception.
Answer B is wrong because a crack in the water jacket in the engine block can cause coolant and oil intermix.
Answer C is wrong because an internal leak in the water pump can cause coolant and engine oil to mix.
Answer D is wrong because a leak in the head gasket can cause coolant intermix.

Question #116
Answer A is wrong because the batteries are connected in parallel to maintain a 12-volt system.
Answer B is correct because the batteries are connected in parallel.
Answer C is wrong because the batteries are connected in parallel to maintain a 12-volt system.
Answer D is wrong because the batteries are connected in parallel to maintain a 12-volt system.

Question #117
Answer A is wrong because the figure shows a stretched valve.
Answer B is wrong because the figure shows a stretched valve.
Answer C is wrong because the figure shows a stretched valve.
Answer D is correct because the figure shows a stretched valve.

Question #118
Answer A is correct because you should first check the bolt holes for foreign matter and cross threading.
Answer B is wrong because the water pump has no effect on head insulation.
Answer C is wrong because the front crank seal is not disturbed when working on a head.
Answer D is wrong because the rear crank seal has no effect on this repair.

Question #119
Answer A is wrong because the cylinder head should be checked for warpage.
Answer B is wrong because the wrist pins are not related to the head.
Answer C is correct.
Answer D is wrong.

Question #120
Answer A is wrong because a cracked fuel pot in the head could cause fuel to leak into the oil.
Answer B is wrong because a broken piston ring could cause fuel to leak into the oil.
Answer C is correct because leaking valve cover gasket is the exception and will not cause fuel to leak into the oil.
Answer D is wrong because a leaking injector seal could cause fuel to leak into the oil.

Question #121
Answer A is wrong because the inside diameter can be checked with a snap gauge.
Answer B is correct because the inside diameter cannot be checked with a feeler gauge.
Answer C is wrong because the inside diameter can be checked with a ball gauge.
Answer D is wrong because the inside diameter can be checked with a dial indicator.

Question #122
Answer A is correct because mechanical rotators on a valve spring will help eliminate carbon deposits on the valve seat.
Answer B is wrong because they cannot prevent oil leakage from the valve guide.
Answer C is wrong.
Answer D is wrong.

Question #123
Answer A is wrong because the cracks allow compression to leak and wear the valve seats.
Answer B is wrong because oblonged seats allow a compression leak and cause poor performance.
Answer C is wrong because discoloration of the back of the valve head will indicate valve seat leakage.
Answer D is correct because worn piston rings will NOT cause a valve seat leak.

Question #124
Answer A is correct because the figure shows the technician replacing valve guide seals.
Answer B is wrong.
Answer C is wrong.
Answer D is wrong.

Question #125
Answer A is a good choice because when removing an injector from a cylinder head you can use a slide hammer to remove the injector. Yet A is wrong because both technicians are right.
Answer B is a good choice because you can also use a puller to remove the injector. Yet B is wrong because both technicians are right.
Answer C is correct.
Answer D is wrong.

Question #126
Answer A is correct because before installing the valves into the cylinder head you should first check valve to seat contact.
Answer B is wrong because the rocker arms are installed after the head is installed.
Answer C is wrong because the lift on the camshaft does not change when replacing valves.
Answer D is wrong because the head gasket installation is a separate process from valve installation.

Question #127
Answer A is wrong because with a wet sleeve type cylinder, the cylinder only seals at the top and bottom.
Answer B is correct because in this type of liner the coolant can pass around the cylinder.
Answer C is wrong.
Answer D is wrong.

Question #128
Answer A is wrong because you check the mating surfaces between the sleeve and the cylinder block.
Answer B is wrong because you do check the mating surfaces between the sleeve and O-rings.
Answer C is wrong because you do check the inside of the sleeve for scoring.
Answer D is correct because you do not check the clearance between the crankshaft and the main caps.

Question #129
Answer A is wrong because the camshaft has no direct effect on piston removal.
Answer B is wrong because the water pump has no direct effect on piston removal.
Answer C is correct because you must remove the top ridge on the cylinder or sleeve first.
Answer D is wrong because the water pump has no direct effect on piston removal.

Question #130
Answer A is wrong because the technician is measuring the cam journal, not the cam lobe.
Answer B is correct because the technician is measuring the cam journal.
Answer C is wrong because the technician is measuring the cam journal.
Answer D is wrong because cam duration has no connection to the figure.

Question #131
Answer A is correct because a micrometer is used to measure the crankshaft journals.
Answer B is wrong because when cleaning the journals you never use a wire wheel or a die grinder.
Answer C is wrong because only one technician is right.
Answer D is wrong because only one technician is right.

Question #132
Answer A is wrong because flywheel housing bore radial runout is being shown in the figure.
Answer B is wrong because flywheel housing bore radial runout is being shown in the figure.
Answer C is wrong because flywheel housing bore radial runout is being shown in the figure.
Answer D is correct because housing bore radial runout is being shown in the figure.

Question #133
Answer A is wrong because a dial indicator is most commonly used to check backlash in the timing gear train.
Answer B is also wrong because a feeler gauge can be used to measure backlash.
Answer C is correct because the LEAST used tool for this process is the depth gauge.
Answer D is wrong because a ruler can be used to check backlash.

Question #134
Answer A is correct because the governor affects a low power complaint the least.
Answer B is wrong because the high idle overrun can be affected by governor operation.
Answer C is wrong because low idle underrun can be affected by governor operation.
Answer D is wrong because a rough running engine is a symptom of low power.

Question #135
Answer A is a good choice because it is easier to install piston rings on an aluminum piston if you preheat the piston in 200°F (93°C) water. Yet A is wrong because both technicians are right.
Answer B is a good choice because the ring ends should be staggered. Yet B is wrong because both technicians are right.
Answer C is correct.
Answer D is wrong.

Question #136
Answer A is wrong.
Answer B is correct because a mechanical governor is shown.
Answer C is wrong.
Answer D is wrong.

Question #137
Answer A is wrong.
Answer B is wrong.
Answer C is wrong.
Answer D is correct because you need to mark all components in the drive train for the purpose of correct balance.

Question #138
Answer A is correct because the crankshaft journals should be machined to fit the bearings.
Answer B is wrong because the main caps should not always be line bored.
Answer C is wrong because only one technician is right.
Answer D is wrong because one technician is right.

Question #139
Answer A is wrong because low voltage will cause false readings.
Answer B is wrong because shorted or open wires will cause false readings.
Answer C is correct because low coolant level has little effect on oil pressure.
Answer D is wrong because a defective sending unit will cause false readings.

Question #140
Answer A is correct because an external oil pump is shown.
Answer B is wrong because an internal pump is mounted in the engine oil pan.
Answer C is wrong because there is no known application of an electric engine oil pump.
Answer D is wrong because there is no known application of a pneumatic engine oil pump.

Question #141
Answer A is a good choice because there are two different types of filters used, the cartridge and the spin-on type. Yet A is wrong because both technicians are right.
Answer B is a good choice because they also use a partial-type and a full-flow type. Yet B is wrong because both technicians are right.
Answer C is correct.
Answer D is wrong.

Question #142
Answer A is wrong because the part shown in the figure is the oil cooler.
Answer B is correct because the part shown in the figure is the oil cooler.
Answer C is wrong because the part shown in the figure is the oil cooler.
Answer D is wrong because the part shown in the figure is the oil cooler.

Question #143
Answer A is wrong because you place heat sensor tape on the radiator upper tank.
Answer B is wrong because you place the thermometer in the radiator after the engine has reached normal operating temperature.
Answer C is wrong because you point the heat gun at the upper radiator hose.
Answer D is correct because voltmeters are seldom used to check temperature gauges.

Question #144
Answer A is correct because a stuck thermostat could cause this problem.
Answer B is wrong because an overfilled radiator could NOT cause this problem.
Answer C is wrong because only one technician is right.
Answer D is wrong because one technician is right.

Question #145
Answer A is correct because the process shown in the figure is flywheel clutch face surface runout.
Answer B is wrong because the process shown in the figure is flywheel clutch face surface runout.
Answer C is wrong because the process shown in the figure is flywheel clutch face surface runout.
Answer D is wrong because the process shown in the figure is flywheel clutch face surface runout.

Question #146
Answer A is wrong because the starter ring gear is never pinned or bolted to the flywheel assembly.
Answer B is correct because the ring gear is a shrink fit to the flywheel.
Answer C is wrong because only one technician is right.
Answer D is wrong because one technician is right.

Question #147
Answer A is a good choice because when installing an in-line injector pump with an electronic governor, it must be timed with the engine camshaft to ensure proper injection timing. Yet A is wrong because both technicians are right.
Answer B is a good choice because an internal camshaft in the pump determines the firing order. Yet B is wrong because both technicians are right.
Answer C is correct.
Answer D is wrong.

Question #148
Answer A is wrong because a plugged fuel line will cut off the fuel supply to the pump.
Answer B is wrong because a low idle will cause the pump to run slower causing lower fuel pressure.
Answer C is wrong because an injector that is sticking open will cause the pressure from the pump to be lower.
Answer D is correct because low-coolant level is the LEAST likely cause of low-fuel pressure.

Question #149
Answer A is a good choice because high-pressure fuel lines should be bled of any air that might be trapped after replacement. Yet A is wrong because both technicians are right.
Answer B is a good choice because you can check for air in the fuel system by installing a clear plastic hose and look for large bubbles. Yet B is wrong because both technicians are right.
Answer C is correct.
Answer D is wrong.

Question #150
Answer A is wrong because an oil bath air cleaner forces air through the oil in the base.
Answer B is correct because a dry-type air cleaner is shown in the figure.
Answer C is wrong because the figure shows a dry-type air cleaner.
Answer D is wrong because the figure shows a dry-type air cleaner.

Question #151
Answer A is wrong because both the air intake and air filter are connected.
Answer B is wrong because the exhaust manifold is connected to the turbocharger.
Answer C is wrong because the oil feed line is connected to the turbocharger.
Answer D is correct because the water pump is not connected to the turbocharger.

Question #152
Answer A is wrong because if the blower bearings leak it could cause this condition.
Answer B is wrong because the rotor bearings are fed by engine oil.
Answer C is correct because a restriction in the oil feed line is the LEAST likely cause of film on the rotor.
Answer D is wrong because a worn end plate gasket could cause this conditon.

Question #153
Answer A is a good choice because an exhaust leak will effect the performance of the engine. Yet A is wrong because both technicians are right.
Answer B is a good choice because leaks in the intake pipe can cause damage to the intake side of the turbo. Yet B is wrong because both technicians are right.
Answer C is correct.
Answer D is wrong.

Question #154
Answer A is correct because the technician should first check the intercooler when he or she notices coolant in the intake side of the turbocharger.
Answer B is wrong because the water is not directly connected to the turbocharger.
Answer C is wrong because the intake manifold gasket is connected after the turbocharger.
Answer D is wrong because the turbocharger is not directly connected to the head.

Question #155
Answer A is wrong because in the figure the technician is checking turbine shaft end play.
Answer B is correct because the figure shows the checking of the turbocharger turbine shaft end play.
Answer C is wrong because in the figure the technician is checking turbine shaft end play.
Answer D is wrong because in the figure the technician is checking turbine shaft end play.

Question #156
Answer A is wrong because a compression brake will NOT cause the engine to stop the vehicle immediately.
Answer B is correct because the compression brake opens the exhaust valves before the compression stroke is complete. Yet B is wrong because both technicians are right.
Answer C is wrong.
Answer D is wrong.

Question #157
Answer A is wrong because the master piston is part of the system.
Answer B is wrong because the slave piston is part of the system.
Answer C is wrong because the solenoid valve is part of the system.
Answer D is correct because brake shoes are not part of a compression brake.

Question #158
Answer A is correct because you should check the air cleaner first.
Answer B is wrong.
Answer C is wrong.
Answer D is wrong.

Question #159
Answer A is wrong because low air box pressure will NOT cause an air inlet restriction.
Answer B is correct because you can check air box pressure with a water manometer.
Answer C is wrong because only one technician is right.
Answer D is wrong because one technician is right.

Question #160
Answer A is correct because on a vehicle with an electric fuel shut-off valve the oil pressure gauge is not part of the system.
Answer B is wrong because the key switch is part of the system.
Answer C is wrong because the solenoid is part of the system.
Answer D is wrong because the governor linkage is part of the system.

Question #161
Answer A is wrong because the wrong muffler will cause excessive backpressure.
Answer B is wrong because excessive carbon will cause excessive backpressure.
Answer C is wrong because a collapsed exhaust pipe will cause excessive backpressure.
Answer D is correct because low oil pressure is least likely to cause high exhaust backpressure.

Question #162
Answer A is correct because excessive crankcase pressure can be caused from a plugged oil breather.
Answer B is wrong because only some diesel engines use EGR valves.
Answer C is wrong because only one technician is right.
Answer D is wrong because one technician is right.

Question #163
Answer A is wrong.
Answer B is correct because electrical current must be disconnected from all sensors at the same time.
Answer C is wrong.
Answer D is wrong.

Question #164
Answer A is wrong because any kinks in the lines will cause low-fuel pressure at the injectors or injection pump.
Answer B is wrong because corroded fuel connectors can cause low-fuel pressure at the injectors or injection pump.
Answer C is wrong because you should check the lines.
Answer D is correct because the power steering pump is the exception and will not cause low-fuel pressure at the injectors or injection pump.

Question #165
Answer A is wrong because if an engine has low horsepower, the fuel filter will not have a direct effect on horsepower.
Answer B is wrong because the high idle setting will not affect horsepower.
Answer C is wrong.
Answer D is correct.

Question #166
Answer A is a good choice because the powertrain control module (PCM) contains emission-related information codes. Yet A is wrong because both technicians are right.
Answer B is a good choice because you must have a scan tool to clear the emission codes. Yet B is wrong because both technicians are right.
Answer C is correct.
Answer D is wrong.

Question #167
Answer A is wrong.
Answer B is correct because DTC stands for diagnostic trouble codes.
Answer C is wrong.
Answer D is wrong.

Question #168
Answer A is wrong because you do not have to remove the battery.
Answer B is wrong because the starter solenoid is removed with the starter.
Answer C is wrong because the voltage regulator is not part of the starter.
Answer D is correct because you must disconnect the battery cables first.

Question #169
Answer A is wrong because the figure shows positive and negative battery plates.
Answer B is correct because the figure shows positive and negative battery plates.
Answer C is wrong because the figure shows positive and negative battery plates.
Answer D is wrong because the figure shows positive and negative battery plates.

Question #170
Answer A is a good choice because using a pressure cap on the radiator, the boiling point of the water will be raised. Yet A is wrong because both technicians are right.
Answer B is a good choice because pressure caps come with many different pressure settings. Yet B is wrong because both technicians are right.
Answer C is correct.
Answer D is wrong.

Question #171
Answer A is wrong.
Answer B is correct because you bleed the injector line first.
Answer C is wrong.
Answer D is wrong.

Question #172
Answer A is correct because when performing a starter draw test you should not crank the engine for more than 30 seconds at a time.
Answer B is wrong because you cannot perform this test with a battery charger connected to the battery because then you would not be testing just the battery capacity.
Answer C is wrong because only one technician is right.
Answer D is wrong because one technician is right.

Question #173
Answer A is wrong because a battery that shows readings below specifications should be replaced.
Answer B is wrong because a battery that shows readings below specifications should be replaced.
Answer C is correct because you need to replace the battery.
Answer D is wrong because voltage regulation has no effect on the battery during a load test.

Question #174
Answer A is correct because radiator air shutters are used to help maintain coolant temperature.
Answer B is wrong because air shutters are not used to keep out foreign debris from the radiator.
Answer C is wrong because only one technician is right.
Answer D is wrong because one technician is right.

Question #175
Answer A is wrong.
Answer B is wrong.
Answer C is correct because the figure shows a thermo-modulated fan hub.
Answer D is wrong.

Question #176
Answer A is correct because glow plugs are used to help start the engine in cold weather.
Answer B is wrong because glow plugs should NOT be disconnected in warm weather.
Answer C is wrong because only one technician is right.
Answer D is wrong because one technician is right.

Question #177
Answer A is wrong because the pump is not part of the system.
Answer B is wrong because the inlet check valve is not part of the system.
Answer C is wrong because the outlet check valve is not part of the system.
Answer D is correct because the fan pulley is part of the system.

Question #178
Answer A is wrong because four valve heads on a four stroke engine use two intake valves and two exhaust valves in each cylinder.
Answer B is correct because a four stroke diesel engine with valve crossheads will use a four valve head.
Answer C is wrong.
Answer D is wrong.

Question #179
Answer A is wrong because the injector lines should be bled of all air.
Answer B is wrong because the primer pump can be used to bleed the injector lines.
Answer C is correct because replacing the fuel filter is not necessary in the bleeding process.
Answer D is wrong because pressurizing the tank with a small amount of air pressure is acceptable.

Question #180
Answer A is wrong.
Answer B is correct because the technician is replacing the valve guide.
Answer C is wrong.
Answer D is wrong.

Question #181
Answer A is wrong because the air shut down valve can be mounted to the blower housing.
Answer B is wrong because the air shut down valve can be mounted to the turbo inlet side.
Answer C is wrong because the air shut down valve can be mounted to the housing.
Answer D is correct because the air shut down valve is least likely to be mounted to the intake manifold.

Question #182
Answer A is correct because the head rail is not part of the valve cover assembly.
Answer B is wrong because the breather can be mounted on top of the valve cover.
Answer C is wrong because the filter cap is usually part of the valve cover.
Answer D is wrong because the strainer is mounted under the breather.

Question #183
Answer A is a good choice because a shutoff solenoid has two coils in one housing. Yet A is wrong because both technicians are right.
Answer B is a good choice because one coil pulls the fuel shutoff to the "open" position. Yet B is wrong because both technicians are right.
Answer C is correct.
Answer D is wrong.

Question #184
Answer A is wrong because the block should always be checked for cracks first before any assembly.
Answer B is correct because the block should always be checked for cracks first before any assembly.
Answer C is wrong because the sleeves should be removed before cleaning the block.
Answer D is wrong because the block should always be checked for cracks first before any assembly.

Question #185
Answer A is correct because when replacing a wet-type sleeve you should always replace the seals on the sleeve.
Answer B is wrong because to remove the sleeve you use a sleeve puller, you do not tap it out with a hammer.
Answer C is wrong.
Answer D is wrong.

Question #186
Answer A is wrong because a digital ohmmeter can be used to test connections on PCM.
Answer B is wrong because a DMM can be used to test connections on PCM.
Answer C is wrong because a high impedance test light can be used to test connections on PCM.
Answer D is correct because a load tester is not used to test connections on PCM.

Question #187
Answer A is a good choice because the customer option codes can be changed in the electronic systems with a scan tool. Yet A is wrong because both technicians are right.
Answer B is a good choice because in some cases changing these codes could result in poor engine performance. Yet B is wrong because both technicians are right.
Answer C is correct.
Answer D is wrong.

Question #188
Answer A is wrong because these numbers are used to show the tightening sequence of the heads bolts.
Answer B is correct because it tells the technician the tightening order to provide the proper clamping load on the head.
Answer C is wrong because these numbers are used to show the tightening sequence of the heads bolts.
Answer D is wrong because these numbers are used to show the tightening sequence of the heads bolts.

Question #189
Answer A is a good choice because the throttle position sensor is connected to a 5-volt reference. Yet A is wrong because both technicians are right.
Answer B is a good choice because when the throttle is manually opened, the scan tool should read a low voltage. Yet B is wrong because both technicians are right.
Answer C is correct.
Answer D is wrong.

Question #190
Answer A is wrong because the main bearing caps do not have to be removed to inspect the block bores.
Answer B is correct because the block bores must be inspected.
Answer C is wrong because the oil ports can be inspected after inspecting the block.
Answer D is wrong because the cam bearings are not directly connected to the cylinder liners.

Question #191
Answer A is a good choice because the EMC connectors should never be probed through the weather pack seals. Yet A is wrong because both technicians are right.
Answer B is a good choice because a test kit should be used to test connectors. Yet B is wrong because both technicians are right.
Answer C is correct.
Answer D is wrong.

Question #192
Answer A is wrong because the solenoid is located inside the pump.
Answer B is correct because you remove the pump cover to replace the solenoid.
Answer C is wrong because the injection pump does not have to be removed to remove the solenoid.
Answer D is wrong because the solenoid is located inside the pump.

Question #193
Answer A is correct because cylinder heads are always torque in sequence according to specifications set by the manufacturer.
Answer B is wrong because bent push rods must always be replaced.
Answer C is wrong.
Answer D is wrong.

Question #194
Answer A is a good choice because it is easier to install piston rings on an aluminum piston if you preheat the piston in 200°F (93°C) water. Yet A is wrong because both technicians are right.
Answer B is a good choice because the ring ends should be staggered. Yet B is wrong because both technicians are right.
Answer C is correct.
Answer D is wrong because neither technician is right.

Question #195
Answer A is correct because adding or removing shims moves the rotor lobs closer or farther apart, adjusting the clearance.
Answer B is wrong because changing the end cover will not adjust clearance.
Answer C is wrong because using over or under sized gears will not affect clearance and these parts do not exist.
Answer D is wrong because changing bearings is the appropriate process.

Question #196
Answer A is wrong because the timing gears should be replaced if they show signs of wear.
Answer B is correct because the lifter should be checked in the mating bore in the block to ensure proper fit.
Answer C is wrong.
Answer D is wrong.

Question #197
Answer A is wrong because the pulleys should only be replaced if they are worn.
Answer B is wrong because the idler pulleys should only be replaced if they are worn.
Answer C is correct because you check belts for cracking and wear.
Answer D is wrong because the belt dressing should be applied after adjusting the belts.

Question #198
Answer A is wrong.
Answer B is wrong.
Answer C is wrong.
Answer D is correct because the technician is checking the spray pattern of the injector.

Glossary

ABS An abbreviation for Anti-lock Brake System.

Absolute Pressure The aero point from which pressure is measured.

Ackerman Principle The geometric principle used to provide toe-out on turns. The ends of the steering arms are angled so that the inside wheel turns more than the outside wheel when a vehicle is making a turn.

Actuator A device that delivers motion in response to an electrical signal.

Adapter The welds under a spring seat to increase the mounting height or fit a seal to the axle.

Adapter Ring A part that is bolted between the clutch cover and the flywheel on some two-plate clutches when the clutch is installed on a flat flywheel.

A/D Converter An abbreviation for Analog-to-Digital Converter.

Additive An additive intended to improve a certain characteristic of the material.

Adjustable Torque Arm A member used to retain axle alignment and, in some cases, control axle torque. Normally one adjustable and one rigid torque arm are used per axle so the axle can be aligned. This rod has means by which it can be extended or retracted for adjustment purposes.

Adjusting Ring A device that is held in the shift signal valve bore by a press fit pin through the valve body housing. When the ring is pushed in by the adjusting tool, the slots on the ring that engage the pin are released.

After-Cooler A device that removes water and oil from the air by a cooling process. The air leaving an after-cooler is saturated with water vapor, which condenses when a drop in temperature occurs.

Air Bag An air-filled device that functions as the spring on axles that utilize air pressure in the suspension system.

Air Brakes A braking system that uses air pressure to actuate the brakes by means of diaphragms, wedges, or cams.

Air Brake System A system utilizing compressed air to activate the brakes.

Air Compressor (1) An engine-driven mechanism for supplying high pressure air to the truck brake system. There are basically two types of compressors: those designed to work on in-line engines and those that work on V-type engines. The in-line type is mounted forward and is gear driven, while the V-type is mounted toward the firewall and is camshaft driven. With both types the coolant and lubricant are supplied by the truck engine. (2) A pump-like device in the air conditioning system that compresses refrigerant vapor to achieve a change in state for the refrigeration process.

Air Conditioning The control of air movement, humidity, and temperature by mechanical means.

Air Dryer A unit that removes moisture.

Air Filter/Regulator Assembly A device that minimizes the possibility of moisture-laden air or impurities entering a system.

Air Hose An air line, such as one between the tractor and trailer, that supplies air for the trailer brakes.

Air-Over-Hydraulic Brakes A brake system utilizing a hydraulic system assisted by an air pressure system.

Air-Over-Hydraulic Intensifier A device that changes the pneumatic air pressure from the treadle brake valve into hydraulic pressure which controls the wheel cylinders.

Air Shifting The process that uses air pressure to engage different range combinations in the transmission's auxiliary section without a mechanical linkage to the driver.

Air Slide Release A release mechanism for a sliding fifth wheel, which is operated from the cab of a tractor by actuating an air control valve. When actuated, the valve energizes an air cylinder, which releases the slide lock and permits positioning of the fifth wheel.

Air Spring An airfilled device that functions as the spring on axles that utilize air pressure in the suspension system.

Air Spring Suspension A single or multi-axle suspension relying on air bags for springs and weight distribution of axles.

Air Timing The time required for the air to be transmitted to or released from each brake, starting the instant the driver moves the brake pedal.

Altitude Compensation System An altitude barometric switch and solenoid used to provide better driveability at more than 4,000 feet (1220 meters) above sea level.

Ambient Temperature Temperature of the surrounding or prevailing air. Normally, it is considered to be the temperature in the service area where testing is taking place.

Amboid Gear A gear that is similar to the hypoid type with one exception: the axis of the drive pinion gear is located above the centerline axis of the ring gear.

Amp An abbreviation for ampere.

Ampere The unit for measuring electrical current.

Analog Signal A voltage signal that varies within a given range (from high to low, including all points in between).

Analog-to-Digital Converter (A/D converter) A device that converts analog voltage signals to a digital format; this is located in a section of the processor called the input signal conditioner.

Analog Volt/Ohmmeter (AVOM) A test meter used for checking voltage and resistance. Analog meters should not be used on solid state circuits.

Annulus The largest part of a simple gear set.

Anticorrosion Agent A chemical used to protect metal surfaces from corrosion.

Antifreeze A compound, such as alcohol or glycerin, that is added to water to lower its freezing point.

Anti-lock Brake System (ABS) A computer controlled brake system having a series of sensing devices at each wheel that control braking action to prevent wheel lockup.

Anti-lock Relay Valve (ARV) In an anti-lock brake system, the device that usually replaces the standard relay valve used to control the rear axle service brakes and performs the standard relay function during tractor/trailer operation.

Antirattle Springs Springs that reduce wear between the intermediate plate and the drive pin, and helps to improve clutch release.

Antirust Agent An additive used with lubricating oils to prevent rusting of metal parts when the engine is not in use.

Application Valve A foot-operated brake valve that controls air pressure to the service chambers.

Applied Moment A term meaning a given load has been placed on a frame at a particular point.

Area The total cross section of a frame rail including all applicable elements usually given in square inches.

Armature The rotating component of a (1) starter or other motor. (2) generator. (3) compressor clutch.

Articulating Upper Coupler A bolster plate kingpin arrangement that is not rigidly attached to the trailer, but provides articulation and/or oscillation, (such as a frameless dump) about an axis parallel to the rear axle of the trailer.

Articulation Vertical movement of the front driving or rear axle relative to the frame of the vehicle to which they are attached.

ASE An abbreviation for Automotive Service Excellence, a trademark of National Institute for Automotive Service Excellence.

Aspect Ratio A tire term calculated by dividing the tire's section height by its section width.

ATEC System A system that includes an electronic control system, torque converter, lockup clutch, and planetary gear train.

Atmospheric Pressure The weight of the air at sea level; 14.696 pounds per square inch (psi) or 101.33 kilopascals (kPa).

Automatic Slack Adjuster The device that automatically adjusts the clearance between the brake linings and the brake drum or rotor. The slack adjuster controls the clearance by sensing the length of the stroke of the push rod for the air brake chamber.

Autoshift Finger The device that engages the shift blocks on the yoke bars that corresponds to the tab on the end of the gearshift lever in manual systems.

Auxiliary Filter A device installed in the oil return line between the oil cooler and the transmission to prevent debris from being flushed into the transmission causing a failure. An auxiliary filter must be installed before the vehicle is placed back in service.

Auxiliary Section The section of a transmission where range shifting occurs, housing the auxiliary drive gear, auxiliary main shaft assembly, auxiliary countershaft, and the synchronizer assembly.

Axis of Rotation The center line around which a gear or part revolves.

Axle (1) A rod or bar on which wheels turn. (2) A shaft that transmits driving torque to the wheels.

Axle Range Interlock A feature designed to prevent axle shifting when the interaxle differential is locked out, or when lockout is engaged. The basic shift system operates the same as the standard shift system to shift the axle and engage or disengage the lockout.

Axle Seat A suspension component used to support and locate the spring on an axle.

Axle Shims Thin wedges that may be installed under the leaf springs of single axle vehicles to tilt the axle and correct the U-joint operating angles. Wedges are available in a range of sizes to change pinion angles.

Backing Plate A metal plate that serves as the foundation for the brake shoes and other drum brake hardware.

Battery Terminal A tapered post or threaded studs on top of the battery case, or infernally threaded provisions on the side of the battery for connecting the cables.

Beam Solid Mount Suspension A tandem suspension relying on a pivotal mounted beam, with axles attached at the ends for load equalization. The beam is mounted to a solid center pedestal.

Beam Suspension A tandem suspension relying on a pivotally mounted beam, with axles attached at ends for lead equalization. Beam is mounted to center spring.

Bellows A movable cover or seal that is pleated or folded like an accordion to allow for expansion and contraction.

Bending Moment A term implying that when a load is applied to the frame, it will be distributed across a given section of the frame material.

Bias A tire term where belts and plies are laid diagonally or crisscrossing each other.

Bimetallic Two dissimilar metals joined together that have different bending characteristics when subjected to different changes of temperature.

Blade Fuse A type of fuse having two flat male lugs sticking out for insertion in the female box connectors.

Bleed Air Tanks The process of draining condensation from air tanks to increase air capacity and brake efficiency.

Block Diagnosis Chart A troubleshooting chart that lists symptoms, possible causes, and probable remedies in columns.

Blower Fan A fan that pushes or blows air through a ventilation, heater, or air conditioning system.

Bobtail Proportioning Valve A valve that senses when the tractor is bobtailing and automatically reduces the amount of air pressure that can be applied to the tractor's drive axle(s). This reduces braking force on the drive axles, lessening the chance of a spin out on slippery pavement.

Bobtailing A tractor running without a trailer.

Bogie The axle spring, suspension arrangement on the rear of a tandem axle tractor.

Bolster Plate The flat load-bearing surface under the front of a semitrailer, including the kingpin, which rests firmly on the fifth wheel when coupled.

Bolster Plate Height The height from the ground to the bolster plate when the trailer is level and empty.

Boss A heavy cast section that is used for support, such as the outer race of a bearing.

Bottoming A condition that occurs when; (1) The teeth of one gear touch the lowest point between teeth of a mating gear. (2) The bed or frame of the vehicle strikes the axle, such as may be the case of overloading.

Bottom U-Bolt Plate A plate that is located on the bottom side of the spring or axle and is held in place when the U-bolts are tightened to the clamp spring and axle together.

Bracket An attachment used to secure parts to the body or frame.

Brake Control Valve A dual brake valve that releases air from the service reservoirs to the service lines and brake chambers. The valve includes a piston which pushes on diaphragms to open ports; these vent air to service lines in the primary and secondary systems.

Brake Disc A steel disc used in a braking system with a caliper and pads. When the brakes are applied, the pad on each side of the spinning disc is forced against the disc, thus imparting a braking force. This type of brake is very resistant to brake fade.

Brake Drum A cast metal bell-like cylinder attached to the wheel that is used to house the brake shoes and provide a friction surface for stopping a vehicle.

Brake Fade A condition that occurs when friction surfaces become hot enough to cause the coefficient of friction to drop to a point where the application of severe pedal pressure results in little actual braking.

Brake Lining A special friction material used to line brake shoes or brake pads. It withstands high temperatures and pressure. The molded material is either riveted or bonded to the brake shoe, with a suitable coefficient of friction for stopping a vehicle.

Brake Pad The friction lining and plate assembly that is forced against the rotor to cause braking action in a disc brake system.

Brake Shoe The curved metal part, faced with brake lining, which is forced against the brake drum to produce braking action.

Brake Shoe Rollers A hardware part that attaches to the web of the brake shoes by means of roller retainers. The rollers, in turn, ride on the end of an S-cam.

Brake System The vehicle system that slows or stops a vehicle. A combination of brakes and a control system.

Breakaway Valve A device that automatically seals off the tractor air supply from the trailer air supply when the tractor system pressure drops to 30 or 40 psi (207 to 276 kPa).

British Thermal Unit (Btu) A measure of heat quantity equal to the amount of heat required to raise 1 pound of water 1°F.

Broken Back Drive Shaft A term often used for non-parallel drive shaft.

Btu An abbreviation for British Thermal Unit.

Bump Steer Erratic steering caused from rolling over bumps, cornering, or heavy braking. Same as orbital steer and roll steer.

CAA An abbreviation for Clean Air Act.

Caliper A disc brake component that changes hydraulic pressure into mechanical force and uses that force to press the brake pads against the rotor and stop the vehicle. Calipers come in three basic types: fixed, floating, and sliding, and can have one or more pistons.

Camber The attitude of a wheel and tire assembly when viewed from the front of a car. If it leans outward, away from the car at the top, the wheel is said to have positive camber. If it leans inward, it is said to have negative camber.

Cam Brakes Brakes that are similar in operation and design to the wedge brake, with the exception that an S-type camshaft is used instead of a wedge and rubber assembly.

Cartridge Fuse A type of fuse having a strip of low melting point metal enclosed in a glass tube. If an excessive current flows through the circuit, the fuse element melts at the narrow portion, opening the circuit and preventing damage.

Caster The angle formed between the kingpin axis and a vertical axis as viewed from the side of the vehicle. Caster is considered positive when the top of the kingpin axis is behind the vertical axis.

Cavitation A condition that causes bubble formation.

Center of Gravity The point around which the weight of a truck is evenly distributed; the point of balance.

Ceramic Fuse A fuse found in some import vehicles that has a ceramic insulator with a conductive metal strip along one side.

CFC An abbreviation for chlorofluorocarbon.

Charging System A system consisting of the battery, alternator, voltage regulator, associated wiring, and the electrical loads of a vehicle. The purpose of the system is to recharge the battery whenever necessary and to provide the current required to power the electrical components.

Charge the Trailer To supply the trailer air tanks with air by means of a dash control valve, tractor protection valve, and a trailer relay emergency valve.

Charging Circuit The alternator (or generator)and associated circuit used to keep the battery charged and to furnish power to the vehicle's electrical systems when the engine is running.

Check Valve A valve that allows air to flow in one direction only. It is a federal requirement to have a check valve between the wet and dry air tanks.

Chlorofluorocarbon (CFC) A compound used in the production of refrigerant that is believed to cause damage to the ozone layer.

Circuit The complete path of an electrical current, including the generating device. When the path is unbroken, the circuit is closed and current flows. When the circuit continuity is broken, the circuit is open and current flow stops.

Clean Air Act (CAA) Federal regulations, passed in 1992, that have resulted in major changes in air-conditioning systems.

Climbing A gear problem caused by excessive wear in gears, bearings, and shafts whereby the gears move sufficiently apart to cause the apex (or point) of the teeth on one gear to climb over the apex of the teeth on another gear with which it is meshed.

Clutch A device for connecting and disconnecting the engine from the transmission or for a similar purpose in other units.

Clutch Brake A circular disc with a friction surface that is mounted on the transmission input spline between the release bearing and the transmission. Its purpose is to slow or stop the transmission input shaft from rotating in order to allow gears to be engaged without clashing or grinding.

Clutch Housing A component that surrounds and protects the clutch and connects the transmission case to the vehicle's engine.

Clutch Pack An assembly of normal clutch plates, friction discs, and one very thick plate known as the pressure plate. The pressure plate has tabs around the outside diameter to mate with the channel in the clutch drum.

COE An abbreviation for cab-over-engine.

Coefficient of Friction A measurement of the amount of friction developed between two objects in physical contact when one of the objects is drawn across the other.

Coil Springs Spring steel spirals that are mounted on control arms or axles to absorb road shock.

Combination A truck coupled to one or more trailers.

Compression Applying pressure to a spring or any springy substance, thus causing it to reduce its length in the direction of the compressing force.

Compressor (1) A mechanical device that increases pressure within a container by pumping air into it. (2) That component of an air-conditioning system that compresses low temperature/pressure refrigerant vapor.

Condensation The process by which gas (or vapor) changes to a liquid.

Condenser A component in an air-conditioning system used to cool a refrigerant below its boiling point causing it to change from a vapor to a liquid.

Conductor Any material that permits the electrical current to flow.

Constant Rate Springs Leaf-type spring assemblies that have a constant rate of deflection.

Control Arm The main link between the vehicle's frame and the wheels that acts as a hinge to allow wheel action up and down independent of the chassis.

Controlled Traction A type of differential that uses a friction plate assembly to transfer drive torque from the vehicle's slipping wheel to the one wheel that has good traction or surface bite.

Converter Dolly An axle, frame, drawbar, and fifth wheel arrangement that converts a semitrailer into a full trailer.

Coolant Liquid that circulates in an engine cooling system.

Coolant Heater A component used to aid engine starting and reduce the wear caused by cold starting.

Coolant Hydrometer A tester designed to measure coolant specific gravity and determine the amount of antifreeze in the coolant.

Cooling System Complete system for circulating coolant.

Coupling Point The point at which the turbine is turning at the same speed as the impeller.

Crankcase The housing within which the crankshaft and many other parts of the engine operate.

Cranking Circuit The starter and its associated circuit, including battery, relay (solenoid), ignition switch, neutral start switch (on vehicles with automatic transmission), and cables and wires.

Cross Groove Joint disc-shaped type of inner CV joint that uses balls and V-shaped grooves on the inner and outer races to accommodate the plunging motion of the half-shaft. The joint usually bolts to a transaxle stub flange; same as disc-type joint.

Cross-Tube A system that transfers the steering motion to the opposite, passenger side steering knuckle. It links the two steering knuckles together and forces them to act in unison.

C-Train A combination of two or more trailers in which the dolly is connected to the trailer by means of two pintle hook or coupler drawbar connections. The resulting connection has one pivot point.

Cycling (1) Repeated on-off action of the air conditioner compressor. (2) Heavy and repeated electrical cycling that can cause the positive plate material to break away from its grids and fall into the sediment chambers at the base of the battery case.

Dampen To slow or reduce oscillations or movement.

Dampened Discs Discs that have dampening springs incorporated into the disc hub. When engine torque is first transmitted to the disc, the plate rotates on the hub, compressing the springs. This action absorbs the shocks and torsional vibration caused by today's low rpm, high torque, engines.

Dash Control Valves A variety of handoperated valves located on the dash. They include parking brake valves, tractor protection valves, and differential lock.

Data Links Circuits through which computers communicate with other electronic devices such as control panels, modules, some sensors, or other computers in the form of digital signals.

Dead Axle Non-live or dead axles are often mounted in lifting suspensions. They hold the axle off the road when the vehicle is traveling empty, and put it on the road when a load is being carried. They are also used as air suspension third axles on heavy straight trucks and are used extensively in eastern states with high axle weight laws. An axle that does not rotate but merely forms a base on which to attach the wheels.

Deadline To take a vehicle out of service.

Deburring To remove sharp edges from a cut.

Dedicated Contract Carriage Trucking operations set up and run according to a specific shipper's needs. In addition to transportation, they often provide other services such as warehousing and logistics planning.

Deflection Bending or moving to a new position as the result of an external force.

Department of Transportation (DOT) A government agency that establishes vehicle standards.

Detergent Additive An additive that helps keep metal surfaces clean and prevents deposits. These additives suspend particles of carbon and oxidized oil in the oil.

DER An abbreviation for Department of Environmental Resources.

Diagnostic Flow Chart A chart that provides a systematic approach to the electrical system and component troubleshooting and repair. They are found in service manuals and are vehicle make and model specific.

Dial Caliper A measuring instrument capable of taking inside, outside, depth, and step measurements.

Differential A gear assembly that transmits power from the drive shaft to the wheels and allows two opposite wheels to turn at different speeds for cornering and traction.

Differential Carrier Assembly An assembly that controls the drive axle operation.

Differential Lock A toggle or push-pull type air switch that locks together the rear axles of a tractor so they pull as one for off-the-road operation.

Digital Binary Signal A signal that has only two values; on and off.

Digital Volt/Ohmmeter (DVOM) A type of test meter recommended by most manufacturers for use on solid state circuits.

Diode The simplest semiconductor device formed by joining P-type semiconductor material with N-type semiconductor material. A diode allows current to flow in one direction, but not in the opposite direction.

Direct Drive The gearing of a transmission so that in its highest gear, one revolution of the engine produces one revolution of the transmission's output shaft. The top gear or final drive ratio of a direct drive transmission would be 1:1.

Disc Brake A steel disc used in a braking system with a caliper and pads. When the brakes are applied, the pad on each side of the spinning disc is forced against the disc, thus imparting a braking force. This type of brake is very resistant to brake fade. A type of brake that generates stopping power by the application of pads against a rotating disc (rotor).

Dispatch Sheet A form used to keep track of dates when the work is to be completed. Some dispatch sheets follow the job through each step of the servicing process.

Dog Tracking Off-center tracking of the rear wheels as related to the front wheels.

DOT An abbreviation for Department of Transportation.

Downshift Control The selection of a lower range to match driving conditions encountered or expected to be encountered. Learning to take advantage of a downshift gives better control on slick or icy roads and on steep downgrades. Downshifting to lower ranges increases engine braking.

Double Reduction Axle An axle that uses two gear sets for greater overall gear reduction and peak torque development. This design is favored for severe service applications, such as dump trucks, cement mixers, and other heavy haulers.

Drag Link A connecting rod or link between the steering gear, Pitman arm, and the steering linkage.

Drawbar Capacity The maximum, horizontal pulling force that can be safely applied to a coupling device.

Driven Gear A gear that is driven or forced to turn by a drive gear, by a shaft, or by some other device.

Drive or Driving Gear A gear that drives another gear or causes another gear to turn.

Drive Line The propeller or drive shaft, universal joints, and so forth, that links the transmission output to the axle pinion gear shaft.

Drive Line Angle The alignment of the transmission output shaft, driveshaft, and rear axle pinion centerline.

Drive Shaft An assembly of one or two universal joints connected to a shaft or tube; used to transmit power from the transmission to the differential.

Drive Train An assembly that includes all power transmitting components from the rear of the engine to the wheels, including clutch/torque converter, transmission, drive line, and front and rear driving axles.

Driver Controlled Main Differential Lock A type of axle assembly has greater flexibility over the standard type of single reduction axle because it provides equal amounts of drive line torque to each driving wheel, regardless of changing road conditions. This design also provides the necessary differential action to the road wheels when the truck is turning a corner.

Driver's Manual A publication that contains information needed by the driver to understand, operate, and care for the vehicle and its components.

Drum Brake A type of brake system in which stopping friction is created by the shoes pressing against the inside of the rotating drum.

Dual Hydraulic Braking System A brake system consisting of a tandem, or double action master cylinder which is basically two master cylinders usually formed by aligning two separate pistons and fluid reservoirs into a single cylinder.

ECU An abbreviation for electronic control unit.

Eddy Current A small circular current produced inside a metal core in the armature of a starter motor. Eddy currents produce heat and are reduced by using a laminated core.

Electricity The movement of electrons from one place to another.

Electric Retarder Electromagnets mounted in a steel frame. Energizing the retarder causes the electromagnets to exert a dragging force on the rotors in the frame and this drag force is transmitted directly to the drive shaft.

Electromotive Force (EMF) The force that moves electrons between atoms. This force is the pressure that exists between the positive and negative points (the electrical imbalance). This force is measured in units called volts.

Electronically Programmable Memory (EPROM) Computer memory that permits adaptation of the ECU to various standard mechanically controlled functions.

Electronic Control Unit (ECU) The brain of the vehicle.

Electronics The technology of controlling electricity.

Electrons Negatively charged particles orbiting around every nucleus.

Elliot Axle A solid bar front axle on which the ends span the steering knuckle.

EMF An abbreviation for electromotive force.

End Yoke The component connected to the output shaft of the transmission to transfer engine torque to the drive shaft.

Engine Brake A hydraulically operated device that converts the vehicle's engine into a power absorbing retarding mechanism.

Engine Stall Point The point, in rpms, under load is compared to the engine manufacturer's specified rpm for the stall test.

Environmental Protection Agency An agency of the United States government charged with the responsibilities of protecting the environment and enforcing the Clean Air Act (CAA) of 1990.

EPA An abbreviation for the Environmental Protection Agency.

EPROM An abbreviation for Electronically Programmable Memory.

Equalizer A suspension device used to transfer and maintain equal load distribution between two or more axles of a suspension. Formerly called a rocker beam.

Equalizer Bracket A bracket for mounting the equalizer beam of a multiple axle spring suspension to a truck or trailer frame while allowing for the beam's pivotal movement. Normally there are three basic types: flange-mount, straddle-mount, and under- or side-mount.

Evaporator A component in an air conditioning system used to remove heat from the air passing through it.

Exhaust Brake A valve in the exhaust pipe between the manifold and the muffler. A slide mechanism which restricts the exhaust flow, causing exhaust back pressure to build up in the engine's cylinders. The exhaust brake actually transforms the engine into a low pressure air compressor driven by the wheels.

External Housing Damper A counterweight attached to an arm on the rear of the transmission extension housing and designed to dampen unwanted driveline or powertrain vibrations.

Extra Capacity A term that generally refers to: (1) A coupling device that has strength capability greater than standard. (2) An oversized tank or reservoir for a fluid or vapor.

False Brinelling The polishing of a surface that is not damaged.

Fanning the Brakes Applying and releasing the brakes in rapid succession on a long downgrade.

Fatigue Failures The progressive destruction of a shaft or gear teeth material usually caused by overloading.

Fault Code A code that is recorded into the computer's memory. A fault code can be read by plugging a special break-out box tester into the computer.

Federal Motor Vehicle Safety Standard (FMVSS) A federal standard that specifies that all vehicles in the United States be assigned a Vehicle Identification Number (VIN).

Federal Motor Vehicle Safety Standard No. 121 (FMVSS 121) A federal standard that made significant changes in the guidelines that cover air brake systems. Generally speaking, the requirements of FMVSS 121 are such that larger capacity brakes and heavier steerable axles are needed to meet them.

FHWA An abbreviation for Federal Highway Administration.

Fiber Composite Springs Springs that are made of fiberglass, laminated, and bonded together by tough polyester resins.

Fifth Wheel A coupling device mounted on a truck and used to connect a semitrailer. It acts as a hinge point to allow changes in direction of travel between the tractor and the semitrailer.

Fifth Wheel Height The distance from the ground to the top of the fifth wheel when it is level and parallel with the ground. It can also refer to the height from the tractor frame to the top of the fifth wheel. The latter definition applies to data given in fifth wheel literature.

Fifth Wheel Top Plate The portion of the fifth wheel assembly that contacts the trailer bolster plate and houses the locking mechanism that connects to the kingpin.

Final Drive The last reduction gear set of a truck.

Fixed Value Resistor An electrical device that is designed to have only one resistance rating, which should not change, for controlling voltage.

Flammable Any material that will easily catch fire or explode.

Flare To spread gradually outward in a bell shape.

Flex Disc A term often used for flex plate.

Flex Plate A component used to mount the torque converter to the crankshaft. The flex plate is positioned between the engine crankshaft and the T/C. The purpose of the flex plate is to transfer crankshaft rotation to the shell of the torque converter assembly.

Float A cruising drive mode in which the throttle setting matches engine speed to road speed, neither accelerating nor decelerating.

Floating Main Shaft The main shaft consisting of a heavy-duty central shaft and several gears that turn freely when not engaged. The main shaft can move to allow for equalization of the loading on the countershafts. This is key to making a twin countershaft transmission workable. When engaged, the floating main shaft transfers torque evenly through its gears to the rest of the transmission and ultimately to the rear axle.

FMVSS An abbreviation for Federal Motor Vehicle Safety Standard.

FMVSS No 121 An abbreviation for Federal Motor Vehicle Safety Standard No 121.

Foot Valve A foot-operated brake valve that controls air pressure to the service chambers.

Foot-Pound An English unit of measurement for torque. One foot-pound is the torque obtained by a force of 1 pound applied to a foot long wrench handle.

Forged Journal Cross Part of a universal joint.

Frame Width The measurement across the outside of the frame rails of a tractor, truck, or trailer.

Franchised Dealership A dealer that has signed a contract with a particular manufacturer to sell and service a particular line of vehicles.

Fretting A result of vibration that the bearing outer race can pick up the machining pattern.

Friction Plate Assembly An assembly consisting of a multiple disc clutch that is designed to slip when a predetermined torque value is reached.

Front Axle Limiting Valve A valve that reduces pressure to the front service chambers, thus eliminating front wheel lockup on wet or icy pavements.

Front Hanger A bracket for mounting the front of the truck or trailer suspensions to the frame. Made to accommodate the end of the spring on spring suspensions. There are four basic types: flange-mount, straddle-mount, under-mount, and side-mount.

Full Trailer A trailer that does not transfer load to the towing vehicle. It employs a tow bar coupled to a swiveling or steerable running gear assembly at the front of the trailer.

Fully Floating Axles An axle configuration whereby the axle half shafts transmit only driving torque to the wheels and not bending and torsional loads that are characteristic of the semi-floating axle.

Fully Oscillating Fifth Wheel A fifth wheel type with fore/aft and side-to-side articulation.

Fusible Link A term often used for fuse link.

Fuse Link A short length of smaller gauge wire installed in a conductor, usually close to the power source.

GCW An abbreviation for gross combination weight.

Gear A disk-like wheel with external or internal teeth that serves to transmit or change motion.

Gear Pitch The number of teeth per given unit of pitch diameter, an important factor in gear design and operation.

General Over-the-Road Use A fifth wheel designed for multiple standard duty highway applications.

Gladhand The connectors between tractor and trailer air lines.

Gross Combination Weight (GCW) The total weight of a fully quipped vehicle including payload, fuel, and driver.

Gross Trailer Weight (GTW) The sum of the weight of an empty trailer and its payload.

Gross Vehicle Weight (GVW) The total weight of a fully equipped vehicle and its payload.

Ground The negatively charged side of a circuit. A ground can be a wire, the negative side of the battery, or the vehicle chassis.

Grounded Circuit A shorted circuit that causes current to return to the battery before it has reached its intended destination.

GTW An abbreviation for gross trailer weight.

GVW An abbreviation for gross vehicle weight.

Halogen Light A lamp having a small quartz/glass bulb that contains a fuel filament surrounded by halogen gas. It is contained within a larger metal reflector and lens element.

Hand Valve (1) A valve mounted on the steering column or dash, used by the driver to apply the trailer brakes independently of the tractor brakes. (2) A hand operated valve used to control the flow of fluid or vapor.

Harness and Harness Connectors The organization of the vehicle's electrical system providing an orderly and convenient starting point for tracking and testing circuits.

Hazardous Materials Any substance that is flammable, explosive, or is known to produce adverse health effects in people or the environment that are exposed to the material during its use.

Heads Up Display (HUD) A technology used in some vehicles that superimposes data on the driver's normal field of vision. The operator can view the information, which appears to "float" just above the hood at a range near the front of a conventional tractor or truck. This allows the driver to monitor conditions such as limited road speed without interrupting his normal view of traffic.

Heater Control Valve A valve that controls the flow of coolant into the heater core from the engine.

Heat Exchanger A device used to transfer heat, such as a radiator or condenser.

Heavy-Duty Truck A truck that has a GVW of 26,001 pounds or more.

Helper Spring An additional spring device that permits greater load on an axle.

High CG Load Any application in which the load center of gravity (CG) of the trailer exceeds 40 inches (102 centimeters) above the top of the fifth wheel.

High-Resistant Circuits Circuits that have an increase in circuit resistance, with a corresponding decrease in current.

High-Strength Steel A low-alloy steel that is much stronger than hot-rolled or cold-rolled sheet steels that normally are used in the manufacture of car body frames.

Hinged Pawl Switch The simplest type of switch; one that makes or breaks the current of a single conductor.

HUD An abbreviation for heads up display.

Hydraulic Brakes Brakes that are actuated by a hydraulic system.

Hydraulic Brake System A system utilizing the properties of fluids under pressure to activate the brakes.

Hydrometer A tester designed to measure the specific gravity of a liquid.

Hypoid Gears Gears that intersect at right angles when meshed. Hypoid gearing uses a modified spiral bevel gear structure that allows several gear teeth to absorb the driving power and allows the gears to run quietly. A hypoid gear is typically found at the drive pinion gear and ring gear interface.

I-Beam Axle An axle designed to give great strength at reasonable weight. The cross section of the axle resembles the letter "I."

ICC Check Valve A valve that allows air to flow in one direction only. It is a federal requirement to have a check valve between the wet and dry air tanks.

Inboard Toward the centerline of the vehicle.

In-Line Fuse A fuse that is in series with the circuit in a small plastic fuse holder, not in the fuse box or panel. It is used, when necessary, as a protection device for a portion of the circuit even though the entire circuit may be protected by a fuse in the fuse box or panel.

In-Phase The in-line relationship between the forward coupling shaft yoke and the driveshaft slip yoke of a two-piece drive line.

Input Retarder A device located between the torque converter housing and the main housing designed primarily for over-the-road operations. The device employs a "paddle wheel" type design with a vaned rotor mounted between stator vanes in the retarder housing.

Installation Templates Drawings supplied by some vehicle manufacturers to allow the technician to correctly install the accessory. The templates available can be used to check clearances or to ease installation.

Insulator A material, such as rubber or glass, that offers high resistance to the flow of electrons.

Integrated Circuit A component containing diodes, transistors, resistors, capacitors, and other electronic components mounted on a single piece of material and capable to perform numerous functions.

Jacobs Engine Brake A term sometimes used for Jake brake.

Jake Brake The Jacobs engine brake, named for its inventor. A hydraulically operated device that converts a power producing diesel engine into a power-absorbing retarder mechanism by altering the engine's exhaust valve opening time used to slow the vehicle.

Jumper Wire A wire used to temporarily bypass a circuit or components for electrical testing. A jumper wire consists of a length of wire with an alligator clip at each end.

Jumpout A condition that occurs when a fully engaged gear and sliding clutch are forced out of engagement.

Jump Start The procedure used when it becomes necessary to use a booster battery to start a vehicle having a discharged battery.

Kinetic Energy Energy in motion.

Kingpin (1) The pin mounted through the center of the trailer upper coupler (bolster plate) that mates with the fifth wheel locks, securing the trailer to the fifth wheel. The configuration is controlled by industry standards. (2) A pin or shaft on which the steering spindle rotates.

Landing Gear The retractable supports for a semitrailer to keep the trailer level when the tractor is detached from it.

Lateral Runout The wobble or side-to-side movement of a rotating wheel or of a rotating wheel and tire assembly.

Lazer Beam Alignment System A two- or four-wheel alignment system using wheel-mounted instruments to project a lazer beam to measure toe, caster, and camber.

Lead The tendency of a car to deviate from a straight path on a level road when there is no pressure on the steering wheel in either direction.

Leaf Springs Strips of steel connected to the chassis and axle to isolate the vehicle from road shock.

Less Than Truckload (LTL) Partial loads from the networks of consolidation centers and satellite terminals.

Light Beam Alignment System An alignment system using wheel-mounted instruments to project light beams onto charts and scales to measure toe, caster, and camber, and note the results of alignment adjustments.

Limited-Slip Differential A differential that utilizes a clutch device to deliver power to either rear wheel when the opposite wheel is spinning.

Linkage A system of rods and levers used to transmit motion or force.

Live Axle An axle on which the wheels are firmly affixed. The axle drives the wheels.

Live Beam Axle A non-independent suspension in which the axle moves with the wheels.

Load Proportioning Valve (LPV) A valve used to redistribute hydraulic pressure to front and rear brakes based on vehicle loads. This is a load- or height-sensing valve that senses the vehicle load and proportions the braking between front and rear brakes in proportion to the load variations and degree of rear-to-front weight transfer during braking.

Lockstrap A manual adjustment mechanism that allows for the adjustment of free travel.

Lock-up Torque Converter A torque converter that eliminates the 10 percent slip that takes place between the impeller and turbine at the coupling stage of operation. It is considered a four-element (impeller, turbine, stator, lockup clutch), three-stage (stall, coupling, and locking stage) unit.

Longitudinal Leaf Spring A leaf spring that is mounted so it is parallel to the length of the vehicle.

Low-Maintenance Battery A conventionally vented, lead/acid battery, requiring normal periodic maintenance.

LTL An abbreviation for less than truckload.

Magnetorque An electromagnetic clutch.

Maintenance-Free Battery A battery that does not require the addition of water during normal service life.

Maintenance Manual A publication containing routine maintenance procedures and intervals for vehicle components and systems.

Main Transmission A transmission consisting of an input shaft, floating main shaft assembly and main drive gears, two counter shaft assemblies, and reverse idler gears.

Manual Slide Release The release mechanism for a sliding fifth wheel, which is operated by hand.

Metering Valve A valve used on vehicles equipped with front disc and rear drum brakes. It improves braking balance during light brake applications by preventing application of the front disc brakes until pressure is built up in the hydraulic system.

Moisture Ejector A valve mounted to the bottom or side of the supply and service reservoirs that collects water and expels it every time the air pressure fluctuates.

Mounting Bracket That portion of the fifth wheel assembly that connects the fifth wheel top plate to the tractor frame or fifth wheel mounting system.

Multiaxle Suspension A suspension consisting of more than three axles.

Multiple Disc Clutch A clutch having a large drum-shaped housing that can be either a separate casting or part of the existing transmission housing.

NATEF An abbreviation for National Automotive Education Foundation.

National Automotive Education Foundation (NATEF) A foundation having a program of certifying secondary and post secondary automotive and heavy-duty truck training programs.

National Institute for Automotive Service Excellence (ASE) A nonprofit organization that has an established certification program for automotive, heavy-duty truck, auto body repair, engine machine shop technicians, and parts specialists.

Needienose Pliers This tool has long tapered jaws for grasping small parts or for reaching into tight spots. Many needlenose pliers also have cutting edges and a wire stripper.

NIASE An abbreviation for National Institute for Automotive Service Excellence, now abbreviated ASE.

NIOSH An abbreviation for National Institute for Occupation Safety and Health.

NLGI An abbreviation for National Lubricating Grease Institute.

NHTSA An abbreviation for National Highway Traffic Safety Administration.

Nonlive Axle Non-live or dead axles are often mounted in lifting suspensions. They hold the axle off the road when the vehicle is traveling empty, and put it on the road when a load is being carried. They are also used as air suspension third axles on heavy straight trucks and are used extensively in eastern states with high axle weight laws.

Nonparallel Driveshaft A type of drive shaft installation whereby the working angles of the joints of a given shaft are equal; however the companion flanges and/or yokes are not parallel.

Nonpolarized Gladhand A gladhand that can be connected to either service or emergency gladhand.

Nose The front of a semitrailer.

No-tilt Convertible Fifth A fifth wheel with fore/aft articulation that can be locked out to produce a rigid top plate for applications that have either rigid and/or articulating upper couplers.

(OEM) An abbreviation for original equipment manufacturer.

Off-road With reference to unpaved, rough, or ungraded terrain on which a vehicle will operate. Any terrain not considered part of the highway system falls into this category.

Ohm A unit of measured electrical resistance.

Ohm's Law The basic law of electricity stating that in any electrical circuit, current, resistance, and pressure work together in a mathematical relationship.

On-road With reference to paved or smooth-graded surface terrain on which a vehicle will operate, generally considered to be part of the public highway system.

Open Circuit An electrical circuit whose path has been interrupted or broken either accidentally (a broken wire) or intentionally (a switch turned off).

Operational Control Valve A valve used to control the flow of compressed air through the brake system.

Oscillation The rotational movement in either fore/aft or side-to-side direction about a pivot point. Generally refers to fifth wheel designs in which fore/aft and side-to-side articulation are provided.

OSHA An abbreviation for Occupational Safety and Health Administration.

Out-of-Phase A condition of the universal joint which acts somewhat like one person snapping a rope held by a person at the opposite end. The result is a violent reaction at the opposite end. If both were to snap the rope at the same time, the resulting waves cancel each other and neither would feel the reaction.

Out-of-Round A wheel or tire defect in which the wheel or tire is not round.

Output Driver An electronic on/off switch that the computer uses to control the ground circuit of a specific actuator. Output drivers are located in the processor along with the input conditioners, microprocessor, and memory.

Output Yoke The component that serves as a connecting link, transferring torque from the transmission's output shaft through the vehicle's drive line to the rear axle.

Oval A condition that occurs when a tube is not round, but is somewhat egg-shaped.

Overall Ratio The ratio of the lowest to the highest forward gear in the transmission.

Overdrive The gearing of a transmission so that in its highest gear one revolution of the engine produces more than one revolution of the transmission's output shaft.

Overrunning Clutch A clutch mechanism that transmits power in one direction only.

Overspeed Governor A governor that shuts off the fuel or stops the engine when excessive speed is reached.

Oxidation Inhibitor (1) An additive used with lubricating oils to keep oil from oxidizing even at very high temperatures. (2) An additive for gasoline to reduce the chemicals in gasoline that react with oxygen.

Pad A disc brake lining and metal back riveted, molded, or bonded together.

Parallel Circuit An electrical circuit that provides two or more paths for the current to flow. Each path has separate resistors and operates independently from the other parallel paths. In a parallel circuit, amperage can flow through more than one resistor at a time.

Parallel Joint Type A type of drive shaft installation whereby all companion flanges and/or yokes in the complete drive line are parallel to each other with the working angles of the joints of a given shaft being equal and opposite.

Parking Brake A mechanically applied brake used to prevent a parked vehicle's movement.

Parts Requisition A form that is used to order new parts, on which the technician writes the names of what part(s) are needed along with the vehicle's VIN or company's identification folder.

Payload The weight of the cargo carried by a truck, not including the weight of the body.

Pipe or Angle Brace Extrusions between opposite hangers on a spring or air-type suspension.

Pitman Arm A steering linkage component that connects the steering gear to the linkage at the left end of the center link.

Pitting Surface irregularities resulting from corrosion.

Planetary Drive A planetary gear reduction set where the sun gear is the drive and the planetary carrier is the output.

Planetary Gear Set A system of gearing that is somewhat like the solar system. A pinion is surrounded by an internal ring gear and planet gears are in mesh between the ring gear and pinion around which all revolve.

Planetary Pinion Gears Small gears fitted into a framework called the planetary carrier.

Plies The layers of rubber-impregnated fabric that make up the body of a tire.

Pogo Stick The air and electrical line support rod mounted behind the cab to keep the lines from dragging between the tractor and trailer.

Polarity The particular state, either positive or negative, with reference to the two poles or to electrification.

Pole The number of input circuits made by an electrical switch.

Pounds per Square Inch (psi) A unit of English measure for pressure.

Power A measure of work being done.

Power Flow The flow of power from the input shaft through one or more sets of gears, or through an automatic transmission to the output shaft.

Power Steering A steering system utilizing hydraulic pressure to reduce the turning effort required of the operator.

Power Synchronizer A device to speed up the rotation of the main section gearing for smoother automatic downshifts and to slow down the rotation of the main section gearing for smoother automatic upshifts.

Power Train An assembly consisting of a drive shaft, coupling, clutch, and transmission differential.

Pressure The amount of force applied to a definite area measured in pounds per square inch (psi) English or kilopascals (kPa) metric.

Pressure Differential The difference in pressure between any two points of a system or a component.

Pressure Relief Valve (1) A valve located on the wet tank, usually preset at 150 psi (1,034 kPa). Limits system pressure if the compressor or governor unloader valve malfunctions. (2) A valve located on the rear head of an air-conditioning compressor or pressure vessel that opens if an excessive system pressure is exceeded.

Printed Circuit Board An electronic circuit board made of thin nonconductive plastic-like material onto which conductive metal, such as copper, has been deposited. Parts of the metal are then etched away by an acid, leaving metal lines that form the conductors for the various circuits on the board. A printed circuit board can hold many complex circuits in a very small area.

Programmable Read Only Memory (PROM) An electronic component that contains program information specific to different vehicle model calibrations.

PROM An abbreviation for Programmable Read Only Memory.

Priority Valve A valve that ensures that the control system upstream from the valve will have sufficient pressure during shifts to perform its automatic functions.

Proportioning Valve A valve used on vehicles equipped with front disc and rear drum brakes. It is installed in the lines to the rear drum brakes, and in a split system, below the pressure differential valve. By reducing pressure to the rear drum brakes, the valve helps to prevent premature lockup during severe brake application and provides better braking balance.

Psi An abbreviation for pounds per square inch.

Pull Circuit A circuit that brings the cab from a fully tilted position up and over the center.

Pull-Type Clutch A type of clutch that does not push the release bearing toward the engine; instead, it pulls the release bearing toward the transmission.

Pump/Impeller Assembly The input (drive) member that receives power from the engine.

Push Circuit A circuit that raises the cab from the lowered position to the desired tilt position.

Push-Type Clutch A type of clutch in which the release bearing is not attached to the clutch cover.

P-type Semiconductors Positively charged materials that enables them to carry current. They are produced by adding an impurity with three electrons in the outer ring (trivalent atoms).

Quick Release Valve A device used to exhaust air as close as possible to the service chambers or spring brakes.

Radial A tire design having cord materials running in a direction from the center point of the tire, usually from bead to bead.

Radial Load A load that is applied at 90° to an axis of rotation.

RAM An abbreviation for random access memory.

Ram Air Air that is forced into the engine or passenger compartment by the forward motion of the vehicle.

Random Access Memory (RAM) The memory used during computer operation to store temporary information. The microcomputer can write, read, and erase information from RAM in any order, which is why it is called random.

Range Shift Cylinder A component located in the auxiliary section of the transmission. This component, when directed by air pressure via low and high ports, shifts between high and low range of gears.

Range Shift Lever A lever located on the shift knob allows the driver to select low or high gear range.

Rated Capacity The maximum, recommended safe load that can be sustained by a component or an assembly without permanent damage.

Ratio Valve A valve used on the front or steering axle of a heavy-duty truck to limit the brake application pressure to the actuators during normal service braking.

RCRA An abbreviation for Resource Conservation and Recovery Act.

Reactivity The characteristic of a material that enables it to react violently with air, heat, water, or other materials.

Read Only Memory (ROM) A type of memory used in microcomputers to store information permanently.

Rear Hanger A bracket for mounting the rear of a truck or trailer suspension to the frame. Made to accommodate the end of the spring on spring suspensions. There are usually four types: flange-mount, straddle-mount, under-mount, and side-mount.

Recall Bulletin A bulletin that pertains to special situations that involve service work or replacement of parts in connection with a recall notice.

Reference Voltage The voltage supplied to a sensor by the computer, which acts as a base line voltage; modified by the sensor to act as an input signal.

Relay An electric switch that allows a small current to control a much larger one. It consists of a control circuit and a power circuit.

Relay/Quick Release Valve A valve used on trucks with a wheel base 254 inches (6.45 meters) or longer. It is attached to an air tank to main supply line to speed the application and release of air to the service chambers. It is similar to a remote control foot valve.

Refrigerant A liquid capable of vaporizing at a low temperature.

Refrigerant Management Center Equipment designed to recover, recycle, and recharge an air-conditioning system.

Release Bearing A unit within the clutch consisting of bearings that mount on the transmission input shaft but do not rotate with it.

Reserve Capacity Rating The ability of a battery to sustain a minimum vehicle electrical load in the event of a charging system failure.

Resistance The opposition to current flow in an electrical circuit.

Resisting Bending Moment A measurement of frame rail strength derived by multiplying the section modulus of the rail by the yield strength of the material. This term is universally used in evaluating frame rail strength.

Resource Conservation and Recovery Act (RCRA) A law that states that after using a hazardous material, it must be properly stored until an approved hazardous waste hauler arrives to take them to the disposal site.

Reverse Elliot Axle A solid-beam front axle on which the steering knuckles span the axle ends.

Revolutions per Minute (rpm) The number of complete turns a member makes in one minute.

Right to Know Law A law passed by the federal government and administered by the Occupational Safety and Health Administration (OSHA) that requires any company that uses or produces hazardous chemicals or substances to inform its employees, customers, and vendors of any potential hazards that may exist in the workplace as a result of using the products.

Rigid Disc A steel plate to which friction linings, or facings, are bonded or riveted.

Rigid Fifth Wheel A platform that is fixed rigidly to a frame. This fifth wheel has no articulation or oscillation. It is generally used in applications where the articulation is provided by other means, such as an articulating upper coupler of a frame-less dump.

Rigid Torque Arm A member used to retain axle alignment and, in some cases, to control axle torque. Normally, one adjustable and one rigid arm are used per axle so the axle can be aligned.

Ring Gear (1) The gear around the edge of a flywheel. (2) A large circular gear such as that found in a final drive assembly.

Rocker Beam A suspension device used to transfer and maintain equal load distribution between two or more axles of a suspension.

Roll Axis The theoretical line that joins the roll center of the front and rear axles.

Roller Clutch A clutch designed with a movable inner race, rollers, accordion (apply) springs, and outer race. Around the inside diameter of the outer race are several cam-shaped pockets. The clutch assembly rollers and accordion springs are located in these pockets.

Rollers A hardware part that attaches to the web of the brake shoes by means of roller retainers. The rollers, in turn, ride on the end of an S-cam.

ROM An abbreviation for read only memory.

Rotary Oil Flow A condition caused by the centrifugal force applied to the fluid as the converter rotates around its axis.

Rotation A term used to describe the fact that a gear, shaft, or other device is turning.

rpm An abbreviation for revolutions per minute.

Rotor (1) A part of the alternator that provides the magnetic fields necessary to create a current flow. (2) The rotating member of an assembly.

Runout A deviation of the specified normal travel of an object. The amount of deviation or wobble a shaft or wheel has as it rotates. Runout is measured with a dial indicator.

Safety Factor (SF) (1) The amount of load which can safely be absorbed by and through the vehicle chassis frame members. (2) The difference between the stated and rated limits of a product, such as a grinding disk.

Screw Pitch Gauge A gauge used to provide a quick and accurate method of checking the threads per inch of a nut or bolt.

Secondary Lock The component or components of a fifth wheel locking mechanism that can be included as a backup system for the primary locks. The secondary lock is not required for the fifth wheel to function and can be either manually or automatically applied. On some designs, the engagement of the secondary lock can only be accomplished if the primary lock is properly engaged.

Section Height The tread center to bead plane on a tire.

Section Width The measurement on a tire from sidewall to sidewall.

Self-Adjusting Clutch A clutch that automatically takes up the slack between the pressure plate and clutch disc as wear occurs.

Semiconductor A solid state device that can function as either a conductor or an insulator, depending on how its structure is arranged.

Semifloating Axle An axle type whereby drive power from the differential is taken by each axle half-shaft and transferred directly to the wheels. A single bearing assembly, located at the outer end of the axle, is used to support the axle half-shaft.

Semioscillating A term that generally describes a fifth wheel type that oscillates or articulates about an axis perpendicular to the vehicle centerline.

Semitrailer A load-carrying vehicle equipped with one or more axles and constructed so that its front end is supported on the fifth wheel of the truck tractor that pulls it.

Sensing Voltage The voltage that allows the regulator to sense and monitor the battery voltage level.

Sensor An electronic device used to monitor relative conditions for computer control requirements.

Series Circuit A circuit that consists of two or more resistors connected to a voltage source with only one path for the electrons to follow.

Series/Parallel Circuit A circuit designed so that both series and parallel combinations exist within the same circuit.

Service Bulletin A publication that provides the latest service tips, field repairs, product improvements, and related information of benefit to service personnel.

Service Manual A manual, published by the manufacturer, that contains service and repair information for all vehicle systems and components.

Shift Bar Housing Available in standard- and forward-position configurations, a component that houses the shift rails, shift yokes, detent balls and springs, inter-lock balls, and pin and neutral shaft.

Shift Fork The Y-shaped component located between the gears on the main shaft that, when actuated, cause the gears to engage or disengage via the sliding clutches. Shift forks are located between low and reverse, first and second, and third and fourth gears.

Shift Rail Shift rails guide the shift forks using a series of grooves, tension balls, and springs to hold the shift forks in gear. The grooves in the forks allow them to interlock the rails, and the transmission cannot be accidentally shifted into two gears at the same time.

Shift Tower The main interface between the driver and the transmission, consisting of a gearshift lever, pivot pin, spring, boot and housing.

Shift Yoke A Y-shaped component located between the gears on the main shaft that, when actuated, cause the gears to engage or disengage via the sliding clutches. Shift yokes are located between low and reverse, first and second, and third and fourth gears.

Shock Absorber A hydraulic device used to dampen vehicle spring oscillations for controlling body sway and wheel bounce, and/or prevent spring breakage.

Short Circuit An undesirable connection between two worn or damaged wires. The short occurs when the insulation is worn between two adjacent wires and the metal in each wire contacts the other, or when the wires are damaged or pinched.

Single-Axle Suspension A suspension with one axle.

Single Reduction Axle Any axle assembly that employs only one gear reduction through its differential carrier assembly.

Slave Valve A valve to help protect gears and components in the transmission's auxiliary section by permitting range shifts to occur only when the transmission's main gearbox is in neutral. Air pressure from a regulator signals the slave valve into operation.

Slide Travel The distance that a sliding fifth wheel is designed to move.

Sliding Fifth Wheel A specialized fifth wheel design that incorporates provisions to readily relocate the kingpin center forward and rearward, which affects the weight distribution on the tractor axles and/or overall length of the tractor and trailer.

Slipout A condition that generally occurs when pulling with full power or decelerating with the load pushing. Tapered or worn clutching teeth will try to "walk" apart as the gears rotate, causing the sliding clutch and gear to slip out of engagement.

Slip Rings and Brushes Components of an alternator that conducts current to the rotor. Most alternators have two slip rings mounted directly on the rotor shaft; they are insulated from the shaft and from each other. A spring loaded carbon brush is located on each slip ring to carry the current to and from the rotor windings.

Solenoid An electromagnet that is used to perform work, made with one or two coil windings wound around an iron tube.

Solid-State Device A device that requires very little power to operate, is very reliable, and generates very little heat.

Solid Wires A single-strand conductor.

Solvent A substance which dissolves other substances.

Spade Fuse A term used for blade fuse.

Spalling Surface fatigue occurs when chips, scales, or flakes of metal break off due to fatigue rather than wear. Spalling is usually found on splines and U-joint bearings.

Specialty Service Shop A shop that specializes in areas such as engine rebuilding, transmission/axle overhauling, brake, air conditioning/heating repairs, and electrical/electronic work.

Specific Gravity The scientific measurement of a liquid based on the ratio of the liquid's mass to an equal volume of distilled water.

Spiral Bevel Gear A gear arrangement that has a drive pinion gear that meshes with the ring gear at the centerline axis of the ring gear. This gearing provides strength and allows for quiet operation.

Splined Yoke A yoke that allows the drive shaft to increase in length to accommodate movements of the drive axles.

Spontaneous Combustion A process by which a combustible material ignites by itself and starts a fire.

Spread Tandem Suspension A two-axle assembly in which the axles are spaced to allow maximum axle loads under existing regulations. The distance is usually more than 55 inches.

Spring A device used to reduce road shocks and transfer loads through suspension components to the frame of the trailer. There are usually four basic types: multileaf, monoleaf, taper, and air springs.

Spring Chair A suspension component used to support and locate the spring on an axle.

Spring Deflection The depression of a trailer suspension when the springs are placed under load.

Spring Rate The load required to deflect the spring a given distance, (usually one inch).

Spring Spacer A riser block often used on top of the spring seat to obtain increased mounting height.

Stability A relative measure of the handling characteristics which provide the desired and safe operation of the vehicle during various maneuvers.

Stabilizer A device used to stabilize a vehicle during turns; sometimes referred to as a sway bar.

Stabilizer Bar A bar that connects the two sides of a suspension so that cornering forces on one wheel are shaped by the other. This helps equalize wheel side loading and reduces the tendency of the vehicle body to roll outward in a turn.

Staff Test A test performed when there is an obvious malfunction in the vehicle's power package (engine and transmission), to determine which of the components is at fault.

Stand Pipe A type of check valve which prevents reverse flow of the hot liquid lubricant generated during operation. When the universal joint is at rest, one or more of the cross ends will be up. Without the stand pipe, lubricant would flow out of the upper passage ways and trunnions, leading to partially dry startup.

Starter Circuit The circuit that carries the high current flow within the system and supplies power for the actual engine cranking.

Starter Motor The device that converts the electrical energy from the battery into mechanical energy for cranking the engine.

Starting Safety Switch A switch that prevents vehicles with automatic transmissions from being started in gear.

Static Balance Balance at rest, or still balance. It is the equal distribution of the weight of the wheel and tire around the axis of rotation so that the wheel assembly has no tendency to rotate by itself regardless of its position.

Stationary Fifth Wheel A fifth wheel whose location on the tractor frame is fixed once it is installed.

Stator A component located between the pump/impeller and turbine to redirect the oil flow from the turbine back into the impeller in the direction of impeller rotation with minimal loss of speed or force.

Stator Assembly The reaction member or torque multiplier supported on a free wheel roller race that is splined to the valve and front support assembly.

Steering Gear A gear set mounted in a housing that is fastened to the lower end of the steering column used to multiply driver turning force and change rotary motion into longitudinal motion.

Steering Stabilizer A shock absorber attached to the steering components to cushion road shock in the steering system, improving driver control in rough terrain and protecting the system.

Stepped Resistor A resistor designed to have two or more fixed values, available by connecting wires to either of the several taps.

Still Balance Balance at rest; the equal distribution of the weight of the wheel and tire around the axis of rotation so that the wheel assembly has no tendency to rotate by itself regardless of its position.

Stoplight Switch A pneumatic switch that actuates the brake lights. There are two types: (1) A service stoplight switch that is located in the service circuit, actuated when the service brakes are applied. (2) An emergency stoplight switch located in the emergency circuit and actuated when a pressure loss occurs.

Storage Battery A battery to provide a source of direct current electricity for both the electrical and electronic systems.

Stranded Wire Wire that is are made up of a number of small solid wires, generally twisted together, to form a single conductor.

Structural Member A primary load-bearing portion of the body structure that affects its over-the-road performance or crash-worthiness.

Sulfation A condition that occurs when sulfate is allowed to remain in the battery plates for a long time, causing two problems: (1) It lowers the specific gravity levels, increasing the danger of freezing at low temperatures. (2) In cold weather a sulfated battery may not have the reserve power needed to crank the engine.

Suspension A system whereby the axle or axles of a unit are attached to the vehicle frame, designed in such a manner that road shocks from the axles are dampened through springs reducing the forces entering the frame.

Suspension Height The distance from a specified point on a vehicle to the road surface when not at curb weight.

Swage To reduce or taper.

Sway Bar A component that connects the two sides of a suspension so that cornering forces on one wheel are shared by the other. This helps equalize wheel side loading and reduces the tendency of the vehicle body to roll outward in a turn.

Switch A device used to control on/off and direct the flow of current in a circuit. A switch can be under the control of the driver or can be self-operating through a condition of the circuit, the vehicle, or the environment.

Synchromesh A mechanism that equalizes the speed of the gears that are clutched together.

Synchro-transmission A transmission with mechanisms for synchronizing the gear speeds so that the gears can be shifted without clashing, thus eliminating the need for double-clutching.

System Protection Valve A valve to protect the brake system against an accidental loss of air pressure, buildup of excess pressure, or back-flow and reverse air flow.

Tachometer An instrument that indicates rotating speeds, sometimes used to indicate crankshaft rpm.

Tag Axle The rearmost axle of a tandem axle tractor used to increase the load-carrying capacity of the vehicle.

Tapped Resistor A resistor designed to have two or more fixed values, available by connecting wires to either of the several taps.

Tandem One directly in front of the other and working together.

Tandem Axle Suspension A suspension system consisting of two axles with a means for equalizing weight between them.

Tandem Drive A two-axle drive combination.

Tandem Drive Axle A type of axle that combines two single axle assemblies through the use of an interaxle differential or power divider and a short shaft that connects the two axles together.

Three-Speed Differential A type of axle in a tandem two-speed axle arrangement with the capability of operating the two drive axles in different speed ranges at the same time. The third speed is actually an intermediate speed between the high and low range.

Throw (1) The offset of a crankshaft. (2) The number of output circuits of a switch.

Tie-Rod Assembly A system that transfers the steering motion to the opposite, passenger side steering knuckle. It links the two steering knuckles together and forces them to act in unison.

Time Guide Prepared reference material used for computing compensation payable by the truck manufacturer for repairs or service work to vehicles under warranty, or for other special conditions authorized by the company.

Timing (1) A procedure of marking the appropriate teeth of a gear set prior to installation and placing them in proper mesh while in the transmission. (2) The combustion spark delivery in relation to the piston position.

Toe A suspension dimension that reflects the difference in the distance between the extreme front and rear of the tire.

Toe In A suspension dimension whereby the front of the tire points inward toward the vehicle.

Toe Out A suspension dimension whereby the front of the tire points outward from the vehicle.

Top U-Bolt Plate A plate located on the top of the spring and is held in place when the U-bolts are tightened to clamp the spring and axle together.

Torque To tighten a fastener to a specific degree of tightness, generally in a given order or pattern if multiple fasteners are involved on a single component.

Torque and Twist A term that generally refers to the forces developed in the trailer and/or tractor frame that are transmitted through the fifth wheel when a rigid trailer, such as a tanker, is required to negotiate bumps, like street curbs.

Torque Converter A component device, similar to a fluid coupling, that transfers engine torque to the transmission input shaft and can multiply engine torque by having one or more stators between the members.

Torque Limiting Clutch Brake A system designed to slip when loads of 20 to 25 pound–feet (27 to 34 N) are reached protecting the brake from overloading and the resulting high heat damage.

Torque Rod Shim A thin wedge-like insert that rotates the axle pinion to change the U-joint operating angle.

Torsional Rigidity A component's ability to remain rigid when subjected to twisting forces.

Torsion Bar Suspension A type of suspension system that utilizes torsion bars in lieu of steel leaf springs or coil springs. The typical torsion bar suspension consists of a torsion bar, front crank, and rear crank with associated brackets, a shackle pin, and assorted bushings and seals.

Total Pedal Travel The complete distance the clutch or brake pedal must move.

Toxicity A statement of how poisonous a substance is.

Tracking The travel of the rear wheels in a parallel path with the front wheels.

Tractor A motor vehicle, without a body, that has a fifth wheel and is used for pulling a semitrailer.

Tractor Protection Valve A device that automatically seals off the tractor air supply from the trailer air supply when the tractor system pressure drops to 30 or 40 psi (207 to 276 kPa).

Tractor/Trailer Lift Suspension A single axle air ride suspension with lift capabilities commonly used with steerable axles for pusher and tag applications.

Trailer A platform or container on wheels pulled by a car, truck, or tractor.

Trailer Hand Control Valve A device located on the dash or steering column and used to apply only the trailer brakes; primarily used in jackknife situations.

Trailer Slider A movable trailer suspension frame that is capable of changing trailer wheelbase by sliding and locking into different positions.

Transfer Case An additional gearbox located between the main transmission and the rear axle to transfer power from the transmission to the front and rear driving axles.

Transistor An electronic device produced by joining three sections of semiconductor materials. Like the diode, it is very useful as a switching device, functioning as either a conductor or an insulator.

Transmission A device used to transmit torque at various ratios and that can usually also change the direction of the force of rotation.

Transverse Vibrations A condition caused by an unbalanced driveline or bending movements, in the drive shaft.

Treadle A dual brake valve that releases air from the service reservoirs to the service lines and brake chambers. The valve includes a piston which pushes on diaphragms to open ports; these vent air to service lines in the primary and secondary systems.

Treadle Valve A foot-operated brake valve that controls air pressure to the service chambers.

Tree Diagnosis Chart A chart used to provide a logical sequence for what should be inspected or tested when troubleshooting a repair problem.

Triaxle Suspension A suspension consisting of three axles with a means of equalizing weight between axles.

Trunnion The end of the universal cross; they are case hardened ground surfaces on which the needle bearings ride.

TTMA An abbreviation for Truck and Trailer Manufacturers Association.

Turbine The output (driven) member that is splined to the forward clutch of the transmission and to the turbine shaft assembly.

TVW An abbreviation for (1) Total vehicle weight. (2) Towed vehicle weight.

Two-Speed Axle Assembly An axle assembly having two different output ratios from the differential. The driver selects the ratios from the controls located in the cab of the truck.

U-Bolt A fastener used to clamp the top U-bolt plate, spring, axle, and bottom U-bolt plate together. Inverted (nuts down) U-bolts cross springs when in place; conventional (nuts up) U-bolts wrap around the axle.

UNEP An abbreviation for United Nations Environment Program. Mandates the complete phaseout of CFC-based refrigerants by 1995.

Underslung Suspension A suspension in which the spring is positioned under the axle.

United Nations Environmental Program (UNEP) A protocol that mandated the complete phase-out of CFC-based refrigerants by the year 1995.

Universal Gladhand A term often used for non-polarized gladhand.

Universal Joint (U-joint) A component that allows torque to be transmitted to components that are operating at different angles.

Upper Coupler The flat load-bearing surface under the front of a semitrailer, including the kingpin, which rests firmly on the fifth wheel when coupled.

Vacuum Air below atmospheric pressure. There are three types of vacuums important to engine and component function: manifold vacuum, ported vacuum, and venturi vacuum. The strength of either of these vacuums depend on throttle opening, engine speed, and load.

Validity List A list supplied by the manufacuirer of valid bulletins.

Valve Body and Governor Test Stand A specialized piece of test equipment. The valve body of the transmission is removed from the vehicle and mounted into the test stand. The test stand duplicates all vehicle running conditions, so the valve body can be thoroughly tested and calibrated.

Variable Pitch Stator A stator design often used in torque converters in off-highway applications such as dirt and stone aggregate dump or haul trucks, or other specialized equipment used to transport unusually heavy loads in rough terrain.

Vehicle Body Clearance (VBC) The distance from the inside of the inner tire to the spring or other body structures.

Vehicle On-board Radar (VORAD) A system similar to an electronic eye that constantly monitors other vehicles on the road to give the driver additional reaction time to respond to potential dangers.

Vehicle Retarder An optional type of braking device that has been developed and successfully used over the years to supplement or assist the service brakes on heavy-duty trucks.

Vertical Load Capacity The maximum, recommended vertical downward force that can be safely applied to a coupling device.

VIN An abbreviation for Vehicle Identification Number.

Viscosity The ability of an oil to maintain proper lubricating quality under various conditions of operating speed, temperature, and pressure. Viscosity describes oil thickness or resistance to flow.

Volt The unit of electromotive force.

Voltage Generating Sensors These are devices which produces their own input voltage signal.

Voltage Limiter A device that provides protection by limiting voltage to the instrument panel gauges to approximately 5 volts.

Voltage Regulator A device that controls the amount of current produced by the alternator or generator and thus the voltage level in the charging circuit.

VORAD An acronym for Vehicle On-board Radar.

Vortex Oil Flow The circular flow that occurs as the oil is forced from the impeller to the turbine and then back to the impeller.

Watt The measure of electrical power.

Watt's Law A basic law of electricity used to find the power of an electrical circuit expressed in watts. It states that power equals the voltage multiplied by the current, in amperes.

Wear Compensator A device mounted in the clutch cover having an actuator arm that fits into a hole in the release sleeve retainer.

Wedge-Actuated Brakes A brake system using air pressure and air brake chambers to push a wedge and roller assembly into an actuator that is located between adjusting and anchor pistons.

Wet Tank A supply reservoir.

Wheel Alignment The mechanics of keeping all the parts of the steering system in the specified relation to each other.

Wheel and Axle Speed Sensors Electromagnetic devices used to monitor vehicle speed information for an antilock controller.

Wheel Balance The equal distribution of weight in a wheel with the tire mounted. It is an important factor which affects tire wear and vehicle control.

Windings (1) The three separate bundles in which wires are grouped in the stator. (2) The coil of wire found in a relay or other similar device. (3) That part of an electrical clutch that provides a magnetic field.

Work (1) Forcing a current through a resistance. (2) The product of a force.

Yield Strength The highest stress a material can stand without permanent deformation or damage, expressed in pounds per square inch (psi).

Yoke Sleeve Kit This can be installed instead of completely replacing the yoke. The sleeve is of heavy walled construction with a hardened steel surface having an outside diameter that is the same as the original yoke diameter.

Zener Diode A variation of the diode, this device functions like a standard diode until a certain voltage is reached. When the voltage level reaches this point, the zener diode will allow current to flow in the reverse direction. Zener diodes are often used in electronic voltage regulators.